D1118997

NUCLEAR POWER

James J. Duderstadt

Chihiro Kikuchi

NUCLEAR POWER

Technology on Trial

Ann Arbor

The University of Michigan Press

Copyright © by The University of Michigan 1979
All rights reserved
Published in the United Sates of America by
The University of Michigan Press and simultaneously
in Rexdale, Canada, by John Wiley & Sons Canada, Limited
Manufactured in the United States of America

1983   1982   1981   1980   5   4

Library of Congress Cataloging in Publication Data

Duderstadt, James J   1942–
  Nuclear power.

  Bibliography: p.
  Includes index.
    1. Atomic power-plants. 2. Atomic power.
I. Kikuchi, Chihiro, 1914–   joint author. II. Title.
TK1078. D83   621.48′3   79–16455
ISBN 0–472–09311–8
ISBN 0–472–06312–X pbk.

To our students

Preface

Nuclear power has become the focus of great controversy. It is almost impossible to pick up a newspaper or turn on the television without encountering some aspect of the debate over this energy source. Are nuclear power plants safe from accident or terrorism? Will they permanently damage man's environment by discharging large quantities of hazardous substances such as radioactive waste or plutonium? Will the international transfer of nuclear power technology accelerate the proliferation of nuclear weapons? Is nuclear power really economical when the staggering increases in plant construction costs, the reliability of nuclear plant operation, and hidden government subsidies are taken into account?

The debate over the future role of nuclear power has taken on particularly ominous overtones in the light of the energy crisis that looms in our immediate future. The current imbalance between our ever-growing energy consumption and our capacity for supplying this energy poses a serious threat to society as we know it. We can achieve a new balance between energy supply and demand only by simultaneously stressing energy conservation while developing new sources of energy as rapidly as possible. Action will be effective only if this nation becomes broadly and pervasively aware of the energy problem, the nature of present energy usage, and the limited availability of conventional energy sources. We must be very realistic in our assessment of the options that are or may become available to alleviate the energy crisis.

As one step to achieve public understanding, we have chosen to analyze the possible role of nuclear power in meeting the future energy requirements of our society. From a more general perspective, nuclear power is an excellent case study of how society accepts or rejects a new technology to meet a perceived need, how it balances the benefits against the risks of the technology to determine its suitability for massive implementation. A detailed study of nuclear power generation exhibits many features that will surface time and time again as our society attempts to adapt technology to its needs. We will evaluate critically the suitability of nuclear power as an energy source by examining this technology in terms of the most significant decision criteria: Does it possess a sufficient resource base? Is its impact on public safety and the environment acceptable? Will it represent an economically viable source of energy?

As the nuclear power industry has matured, as it has passed from Fermi's demonstration of scientific feasibility to its present status of economic viability, so too has grown the public controversy over the role that nuclear power should play in our society. This book introduces readers to the technical issues involved in nuclear power generation so that they can rationally evaluate the future role of this energy source.

We have chosen a level of presentation suitable for the broadest possible audience. No particular scientific background or familiarity has been assumed. Since only a modest introduction to the issues in nuclear power generation can be given in such a brief presentation, we have provided ample documentation to allow the interested reader to pursue further studies.

We must caution the reader that although we have tried to look at both sides of controversial issues, some bias has inevitably crept into our discussion. As scientists and engineers who have participated actively in nuclear energy research for many years, we could hardly be without an opinion. In this spirit we have confined most of our analysis to scientific evidence as we see it. We have avoided merely listing without critical comment the familiar pros and cons of this well-worn debate. We feel that far too few scientists have spoken out on these issues. As a result, the nuclear power debate has frequently drifted away from scientific fact into a dialogue in which scientific misunderstanding and half-truths are rampant. We sincerely hope that more scientific analyses such as this book will correct this alarming trend.

This endeavor has benefited enormously from discussion with and encouragement from a great many colleagues including Professors William Kerr, John Lee, Glenn Knoll, Thomas Brewer, William Martin, David Bach, and John King, along with Drs. Anthony Sinclair, Stanley Borowski, and Robert Campbell. A particular note of gratitude is due Alfred Slote and Anne Duderstadt for their assistance in translating scientific jargon into comprehensible language. We should admit that the real stimulus for this effort came from an entirely different quarter, from those who have been most outspoken in their condemnation of nuclear power. For it was the strong and frequent criticism of all aspects of nuclear power generation voiced by these individuals that stirred within us a strong sense of scientific and social responsibility and that led to this book.

Contents

1

Nuclear Power: Necessary or Not?

Atomic energy! These words conjure up images of the scientists of the Manhattan Project toiling away in secret laboratories to develop terrifying weapons of awesome destruction, images of a mushroom cloud towering above the ruined city of Hiroshima or of nuclear-tipped missiles, poised in underground silos dotting the countryside, awaiting the command that will unleash Armageddon throughout the world. Even today, over a quarter century since the use of the bomb against Japan, atomic energy continues to be regarded as an instrument of war. The nuclear arms race casts an ever-present shadow on international politics and the struggle for world peace.

But the scientists who gave birth to the atomic age realized that the atom had a more benign character and might prove to be the savior rather than the ruin of mankind. Their dream was nuclear power, and they sought to tap the enormous energy contained in the atomic nucleus for the peaceful generation of electricity. Driven by an almost evangelical zeal, these scientists labored to beat their swords into plowshares, to harness atomic energy for the benefit of man.[1] Nuclear power would provide a new source of energy so cheap and abundant that it would not even have to be metered.[2] It could be used to make the deserts bloom, to provide underdeveloped nations with the means to free man once and for all from the ravages of poverty, starvation, and disease. The atom could be used for peace. And by providing man with the energy necessary to fulfill his material desires and eliminate dramatic differences in world living conditions, nuclear power could perhaps eliminate the reasons for war itself.

But the atom was oversold; the excessively optimistic expectations for nuclear power as an instrument of peace were never realized. Much of the early development of nuclear power was characterized by frequent setbacks and frustrations as the new technology fell short of the unrealistic goals that had been set for it.

Some applications, such as the nuclear-powered airplane and the peaceful use of nuclear explosives, were abandoned entirely. And the Dr. Jekyll/Mr. Hyde character of the atom remained as the nuclear arms race escalated. Many who had embraced the early hopes for atomic energy became disenchanted and turned against nuclear technology.

It is ironic that this disenchantment with the atom, stimulated in part by its failure to meet the overly idealistic goals of the postwar world, has occurred just when its most important application, the generation of electric power, has come of age. Through the dedicated efforts of thousands of scientists and engineers and the investment of billions of dollars, nuclear power technology has evolved to the point where it stands ready to meet a significant fraction of the world's requirements for electric energy. Nuclear power plants presently supply roughly 12 percent of the electricity generated in the United States[3] (and a slightly larger percentage in many European nations). Some projections suggest that nuclear power could supply well over half of our electric energy needs by the turn of the century. And it is fortuitous that this has occurred at just that point in history when man is rapidly exhausting his reserves of conventional fuels.

But never before has there been so much controversy over the role that atomic energy is to play in man's future. Just as the fruits of decades of research on nuclear power generation are to be harvested in the form of mammoth nuclear generating stations that can supply mankind's energy needs, there has arisen a very real concern over whether this source of energy should be implemented at all. Are the potential benefits of nuclear power generation worth the risks posed by this technology? Certainly events such as the accident that occurred at the Three Mile Island nuclear power plant in the spring of 1979 cast serious doubts as to the safety of this energy source. Moreover, such plants must have a significant impact on their environment. And what about the dangers and environmental impact of mining uranium ore and processing it into nuclear fuel? What are the dangers of the by-products of nuclear power generation such as plutonium and radioactive wastes? Will the worldwide rush toward nuclear power technology contribute to the spread of nuclear weapons? Are there sufficient uranium resources to sustain nuclear power as a viable, long-term energy source? And would nuclear power truly be economical without massive government subsidy and special intervention, such as the Price-Anderson Act that limits liability in the event of a nuclear accident? What about the unforeseen hazards of future nuclear technology such as the fast breeder reactor or controlled thermonuclear fusion? These, as well as a host of related questions, have been raised in the debate over the future role of nuclear power in our society.

Indeed, from a more general perspective, nuclear power shows us how society accepts or rejects a new technology to meet a perceived need, and how it balances the benefits against the drawbacks of the technology to determine its suitability for massive implementation. As with any new technology, nuclear power had to evolve through a number of stages before it could be applied to the needs of society. As it passed from Fermi's demonstration of scientific feasibility in 1942 through economic viability in the late 1960s toward public acceptance or rejection of its massive implementation during the latter quarter of the twentieth century, nuclear power has been continually on trial. The questions that have arisen during the development of nuclear power technology will arise time and time again as our society attempts to adapt science and technology to its needs. Even today the potential impact of new technologies such as genetic engineering or computer-based data systems is seriously being questioned in a manner reminiscent of the early days of nuclear power development.

Certainly the public trial of a new technology is of immense importance. Society must examine scientific and technological developments carefully to determine the role they should play in meeting its future needs. This evaluation is a dynamic endeavor. It may happen that over the long period of time characterizing the research and development effort required to bring a technology to the state of viability, the criteria used to assess its suitability, and indeed even the original needs that spawned its development, change dramatically.

The public trial of new technologies takes on a particular relevance today because man is pushing against the limits of his environment. The twentieth century marks that point in history at which the overwhelming numbers and appetite of mankind threaten to exhaust the finite resources of this planet.[4] It is important to examine critically the relevance of any new technology for man's future against this ominous backdrop. In many respects the present debate over the future role of nuclear power reflects elements of this concern.

The development of nuclear power has raised many new questions and concerns and stimulated new procedures for evaluating technology. The public has ceased to accept new technologies on the basis of novelty alone and has begun to examine the need for them and their possible impact on public safety and the environment.

Nuclear power particularly has been assessed and regulated in the public spotlight. Indeed, this novel degree of public participation in regulation and assessment has almost certainly contributed to the controversy. For while nuclear power generation is now economically viable and used on a massive scale in most industrialized nations, there is still a bitter debate over whether nuclear technology should be expanded further or phased out and replaced with alternative energy sources.

Perhaps the logical starting point in our examination of the wisdom or folly of committing ourselves to a nuclear future is to determine whether there is really a need for nuclear power. Does our nation or the world need new electric generating capacity? And if so, what alternatives are there to nuclear power development, and what criteria should be used to decide among these alternatives?

Energy Demand versus Supply: The Energy Crisis

There is a serious imbalance between our ever-growing energy consumption and our capacity for supplying this energy. The imbalance is due to many factors, including the energy-intensive nature of our society and way of life, the depletion of existing energy sources (oil and natural gas), and the slow development of new energy sources (advanced schemes for mining and burning coal, nuclear power, solar power, and so on). This imbalance poses a serious threat to our society; the energy crisis is very real. For example, at the present time this country imports almost half of the crude oil it uses, even as it rapidly depletes its own reserves of oil and natural gas.[5] Most serious students of the energy problem agree that a new balance between energy uses and sources of supply must be achieved, and that this can be accomplished only by simultaneously conserving energy and vigorously developing new sources of energy.

Yet the public continues to increase its demand for energy, to purchase and drive large automobiles, to adopt lifestyles based on energy-inefficient devices, to consume materials produced by energy-intensive industrial processes. Interestingly enough, studies[6] have indicated that this continuation of the traditional pattern of rapidly escalating energy consumption is not due to the reluctance of the public to accept a more energy-conservative lifestyle. Rather, the American public does not believe there is an energy crisis. Indeed, disbelief in all societal institutions has reached the point where it is almost impossible for the federal government, or private industry, or the academic community to convince the public of the seriousness of the imbalance between energy demand and production.

For example, prices of oil and natural gas should be readjusted upward to reflect the true replacement costs of these fossil fuels and anticipate their impending scarcity.[7] Yet attempts to do so by price deregulation (as opposed to taxation) inevitably meet cries that such price increases are ploys to increase the profits of the petroleum industry. The same public reaction greets the inevitable increases in electric utility rates as the costs of generating fuels rise. The skepticism that interprets the energy crisis as either a fabrication designed to increase profits or a temporary imbalance between supply and demand is extremely dangerous. Such attitudes destroy the incentive for conservation and the massive commitments necessary for the development of new energy supply technologies. Effective action to alleviate the imbalance can be achieved only if the nation becomes broadly and pervasively aware of the energy problem, the nature of energy usage, and the available resources and their limitations, and rationally approaches the various options available to our society. It is in this light that we must evaluate the possible role that nuclear power development might play in meeting our energy requirements.

Projections of future energy needs determine the magnitude and timing of the investment required to develop new energy supply capacity. Of particular interest are projections concerning future demand for electric energy, since it is this component that will most directly influence the future need for nuclear power generation. We have indicated in figure 1 the historic growth in the demand for energy in the United States.[8] At the present time domestic energy consumption stands at 74 quads per year (a quad is one quadrillion BTUs, the most convenient unit for discussing energy needs and supply). Since 1950, domestic energy consumption has been growing at an average rate of 3.4 percent per year. This increase in energy consumption has followed an exponential rather than a linear growth curve. Our consumption of energy has been accelerating, rather than increasing at a constant rate. This trend has already given rise to an imbalance between the demand for energy and supplies from domestic sources. At the present time the gap between domestic energy supply and demand is being filled by importing some 12 quads per year of foreign crude oil at a cost of some $50 billion.[9]

Future growth in United States energy demand will be determined by a variety of factors: the rate of growth of the economy and the population, changes in demographic and lifestyle factors, and changes in the efficiency of energy use. More specifically, total energy demand can be broken into five components: household, commercial, personal transportation, industry, and transport of goods and services. The first three components are most directly

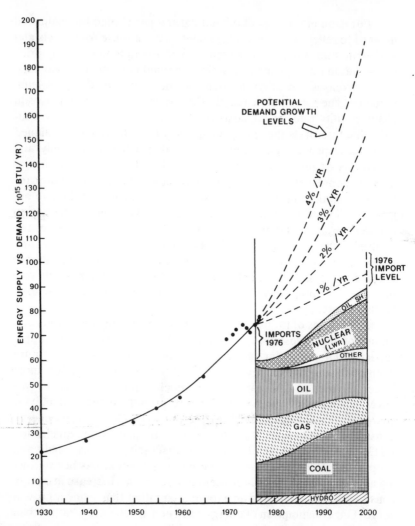

Fig. 1. Growth in United States energy consumption and supply. (*Courtesy of Michigan Energy and Resource Research Association.*)

related to growth in population, while the last two are most closely tied to gross national product (GNP). Therefore, by estimating the growth in population and GNP, one can extrapolate historical energy use in each sector to predict future energy demands.

In recent studies,[10] the Oak Ridge Institute for Energy Analysis relied on relatively modest projections of population growth (ranging between 245 and 254 million by the year 2000) and average increases in GNP of 2.5 to 3.0 percent per year to arrive at an energy demand ranging between 101 and 126 quads per year at the turn of the century (see table 1). This would correspond to a growth rate in energy consumption of some 2.0 to 2.5 percent per year, somewhat below the historical growth rate of 3.4 percent per year. It should be noted that the uncertainty in such projections arises not only from estimates in population and GNP growth, but also from the extrapolation of historical trends that do not anticipate any major changes in lifestyle or technology such as massive conservation efforts or transitions to soft-energy supply sources. However, even these relatively modest growth projections will significantly widen the already serious gap between energy supply and demand.

At the present time some 28 percent of our total energy consumption is in the form of electricity. Most studies project a trend toward a substitution of electricity for other sources of energy to the extent that some 50 percent of our energy consumption will be in electricity by the year 2000.[11] Such a shift would be caused by pressures to replace more expensive fluid fuels (natural gas and oil) with solid fuels (coal and uranium) that are more suited to electric power generation, as well as by advancing technology that relies heavily on electricity as an energy source. Although there have been recent proposals to move away from a society based on central station electric generating systems[12] (proposals that will be discussed in more detail later in this chapter), we favor the projections for future electrical demand of the Institute for Energy Analysis that range between 47 and 64 quads by the turn of the century. This would correspond to an average annual increase in peak load electrical demand of 5 percent (somewhat below the historical growth rate of 7 percent).[13]

Translated into somewhat more dramatic terms, a 5 percent growth rate in electric generating capacity would mean that in a state such as Michigan electric utilities must bring on-line each year a new plant generating 1,000 million watts of electricity (1,000 megawatts-electric).[14] Since such a plant typically costs over $1 billion and requires some ten to twelve years for its construction, the staggering implications of this growth in electrical demand should be readily apparent.

Hence there seems little doubt, at least for the immediate future, that the demand for electric energy will increase. Additional facilities for electric power generation and transmission will be required. Such facilities require long lead times for construction due to the complex nature of the machinery and equipment and the magnitude of the project, as well as the time necessary to obtain permits, approvals, and licenses. These large construction programs involve huge capital outlays that are spread out over a decade. Therefore decisions concerning the type, size, and location of future generating

TABLE 1. Projections of Energy and Electricity Demand

Year	GNP (in billions of 1975$)		Population (in millions)		Energy (in quads)		Electricity (in quads)	
	Low	High	Low	High	Low	High	Low	High
1975	1,499	1,499	213	213	71.1	71.1	20.1	20.1
1985	2,135	2,135	228	231	82.1	88.0	30.8	34.1
2000	3,184	3,326	245	254	101.4	125.9	47.3	64.0
2010	4,076	4,470	250	264	118.3	158.8	55.5	82.4

Source: C. E. Whittle et al., *Economic and Environmental Implications of a United States Nuclear Moratorium, 1985-2010,* Institute for Energy Analysis Report ORAU/IEA 76-4 (Oak Ridge, Tenn., 1976).

facilities must be made today if electric utilities are to meet the electric energy needs of tomorrow.[15]

Decision Criteria for Acceptable Energy Sources

Which types of electric generating facilities are most appropriate? Which energy sources are most capable of meeting anticipated power requirements, while ensuring maximum public safety and minimal environmental impact? To answer such questions, we must consider not only the broad range of alternatives for electric power generation, including fossil fuels (oil, natural gas, and coal), nuclear fission, nuclear fusion, and hydroelectric, geothermal, and solar power, but also what degree of conservation represents a realistic goal.

Certainly the first consideration should be given to the *viability* of the energy alternative during the next ten to twenty years. Here one must take care not to confuse the scientific *feasibility* of a technology, that is, its demonstration on a laboratory scale, with its social viability for massive implementation. For example, although the use of solar photoelectric cells to generate electric power is scientifically feasible (solar power has been used on a small scale for a number of years in space satellites and remote microwave relay stations), there seems to be little chance that the massive implementation of solar photoelectric power generation on an economical basis will be viable until well after the turn of the century. Another example is controlled thermonuclear fusion, which is frequently suggested as an attractive long-range source of energy, but has yet to reach even the stage of scientific feasibility, much less economic viability. Later we shall classify energy alternatives as either short range or long range, depending on whether they will be economically viable by the turn of the century.

A second factor is the *natural resource base* available for a given type of energy source. For example, while oil and natural gas provide over 75 percent of our present energy requirements, their economic viability is expected to diminish as the reserves of these fuels are depleted over the next several decades.[16] In contrast, solar electric power, controlled thermonuclear fusion, and the fast breeder reactor all represent essentially infinite sources of energy, since the estimated resources required for each of these alternatives is either renewable or sufficient to supply mankind's energy needs for thousands (perhaps tens of thousands) of years. Unfortunately, all three of these alternatives must be viewed as long-range options. For the short term, we are forced to look instead to alternatives that exhibit more immediate economic viability but unfortunately rest on more limited fuel resources. This class includes solid fuels such as coal and uranium in addition to options based on rather limited natural conditions such as hydroelectric and geothermal power (see fig. 2).

A third factor that must be considered in any energy program is *public risk*. Certainly any form of electric power generation will entail some degree of risk. The public risks from nuclear power generation (for example, low-level radiation release or accidents such as the Three Mile Island incident) have been widely publicized over the past several years.[17] Not so widely recognized are the rather substantial public risks associated with hydroelectric power generation[18] (dam failures are not uncommon), fires and explosions that invariably occur during the transportation of fluid fossil fuels, and the generation of electricity by burning coal[19] (both from hazards of coal mining and the pollutants emitted by coal-fired generating plants). In fact even soft-energy technologies such as solar power are not entirely risk-free since the industrial processes necessary to fabricate and assemble the enormous quantities of materials necessary to collect, concentrate, and convert solar energy into a useful form entail a significant risk (particularly when viewed on a risk per unit of energy produced basis).[20] It is extremely important in any consideration of various energy alternatives to estimate as realistically as possible the public risk from each alternative and to balance these risks against public benefits.

A fourth area of considerable concern is *environmental impact*. All energy production will disturb the environment to some degree.

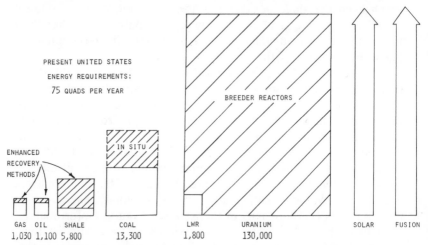

Fig. 2.   Potentially recoverable domestic energy resources.

There is no way to avoid this. The effects of strip-mining coal and uranium are well known. Coal-fired plants discharge enormous quantities of particulate matter and potentially harmful gases such as sulfur dioxide and nitrous oxide directly to the atmosphere, while nuclear plants release radioactivity during normal operation and accumulate intensely radioactive material in the form of spent fuel (radioactive waste) that must be carefully isolated from the environment. Perhaps not so widely recognized are the enormous environmental impacts of seemingly benign sources of energy such as geothermal power (which releases large quantities of noxious gases and liquids to the environment, causes substantial ground settlement and seismic activity, and generates high noise levels) and solar electric power generation (both directly, due to energy collection and storage devices, and indirectly, through the impact of the industrial processes required to manufacture the large quantities of materials necessary for solar power facilities).

Of course, the final and usually deciding consideration will be *economics*. In our society most features of any technology, including resources, safety, and environmental impact, are usually assigned a quantitative measure: the dollar. In the particular case of electric power, the quantitative measure is taken as the cost of generating a certain quantity of electricity, usually expressed in mills per kilowatt-hour, where 1 mill is one tenth of a cent and a kilowatt-hour is a measure of electric energy. In this sense the economic viability or attractiveness of a particular energy option will usually reflect all other aspects involved in selecting that alternative. If our society sets certain minimal levels for environmental impact and public risk, then an energy source must be equipped with sufficient environmental controls and safety systems (usually at a rather significant additional cost) to meet those requirements. The determination of such generating costs is extremely complicated, and the projection of future costs is subject to considerable uncertainty, depending on both timing and location of the projected energy sources.

All of these factors are highly interdependent and frequently conflicting. The efficiency of an energy source may be dramatically affected by environmental controls, such as stack gas scrubbers in coal-fired plants or zero-release radioactivity effluent systems in nuclear plants; hence in many cases, efficiency in power generation can only be purchased at a certain degree of environmental cost. The safety equipment necessary to reduce public risk will lead to higher bills for electricity. There will always be a trade-off.

Ideally all such comparisons of the various factors involved in choosing among energy alternatives, and their advantages and their disadvantages, would be carefully weighed in a rational manner, and an optimum choice would be made. In our society such decisions are frequently made in the most irrational atmospheres, subject to a variety of emotional, political, and economic pressures and conceptual misunderstandings. Much of the confusion in approaching energy issues is due to the numerous agencies—particularly federal agencies—that attempt to regulate and enforce policy decisions involving energy production and utilization. Then, too, there is the inability of this nation to agree on a coordinated energy policy. Unfortunately our society has always had an inclination to seek simple solutions to complex problems, and our efforts to agree on and implement policies that will bring our energy demands and supplies into balance certainly reflects this tendency.

Near-Term Energy Alternatives

What alternatives are available to meet projected demands for energy over the next several decades? Since history has indicated that it takes roughly thirty to forty years to implement a new source of energy, that is, to take it from scientific feasibility to social viability, it is clear that we must temporarily rule out long-range alternatives such as solar electric power and thermonuclear fusion. The only alternatives for the next several decades appear to be a vigorous expansion of the use of solid fuels (coal and uranium) coupled with a massive conservation program. But first let us look at various short-range options.

Fluid Fossil Fuels
Fluid fossil fuels (petroleum and natural gas) resources apparently will supply an appreciable portion of our energy requirements for only a few more decades. Even with additional production from the Alaskan North Slope and continental shelf sites, domestic petroleum production is expected to peak by 1985 and drop off sharply by the end of this century.[21] Worldwide petroleum resources, even when augmented by new discoveries such as those in Mexico's Yucatan Peninsula, are not expected to last beyond the first quarter of the twenty-first century if present consumption trends continue. Hence every attempt is presently being made to shift away from these fuels. However, since over 75 percent of our total energy consumption and some 35 percent of our electric power now comes from this source, our society will remain heavily dependent on these fuels. As our own domestic resources are depleted, we will become increasingly

dependent on the import of foreign petroleum. Indeed, since early 1976, we have been importing almost half of all the crude oil we use (our import of natural gas is still rather small). The rapid increase in the price of foreign crude oil has stimulated price increases in this country to the point where oil-generated electricity is almost three times more expensive than that generated by nuclear plants.[22]

As the price of crude oil continues to rise, it should stimulate the production of oil from shale deposits (at roughly $23 to $26 per barrel projected for 1980) and synthetic oil from coal ($30 per barrel by 1990). In this sense we will not actually run out of fluid fossil fuels by the turn of the century, rather we will be forced to obtain these fuels from new and far more expensive sources. Therefore, although we will continue to use fluid fossil fuels as a major energy source, it would not be prudent to plan on new electric capacity based on gas- or oil-fired plants.

Hydroelectric Power Generation
Certainly hydroelectric power immediately comes to mind when one thinks of the generation of electricity. However, potential hydroelectric sites are rather limited and already have been developed to a large degree. At the present time hydroelectric power accounts for somewhat less than 10 percent of our national electric capacity.[23] Although the projected potential of hydroelectric capacity is roughly twice this, it is rather doubtful that this potential can be achieved because of the serious environmental opposition to hydroelectric development. Furthermore hydroelectric sites are located in only a few regions and certainly would not be appropriate sources for most of the country.

Coal
The coal resources of the United States are quite large.[24] There is little doubt that coal will have the largest impact on our impending energy shortages over the next two decades and that it will continue to supply a large portion of our energy needs throughout the next century. We will expand coal production as rapidly as possible. Most projections anticipate doubling the production of coal over the next decade.[25] However, this increase will require the opening of some four hundred new mines, the employment of an additional 150,000 new miners in underground coal mining, the development of a new rail network, and perhaps the longer range development of slurry pipeline systems to transport coal. Even then the rate at which coal production can be expanded is a subject of rather serious concern.

The occupational hazards of coal mining are well known. Although there is good reason to suspect that suitable federal regulation should improve mine safety, it still may be difficult to attract the necessary labor force to achieve the desired level of coal production. The use of coal also will be limited by a number of environmental concerns.[26] Since our major coal resources are found in the high, arid plateau regions of the west, there is some concern that the water supply may be insufficient for adequate land reclamation, much less coal gasification or liquefaction.[27] For the short term, direct combustion of coal for the generation of electricity seems the most likely use of this fossil fuel. The pollution that will inevitably result from burning large quantities of coal presents a serious hazard to public health. Of most concern are emissions of particulates and gases such as sulfur dioxide and nitrous oxide from coal plants. But there is also considerable concern about the long-term climatic effects of the carbon dioxide buildup in the atmosphere that may result from a major dependence on this energy source.[28]

A long-term commitment will require the development of new technology such as stack gas scrubbing and fluidized-bed combustion for reducing the emissions from coal combustion. Already the Environmental Protection Agency (EPA) has ruled that all new coal-fired generating plants must be equipped with stack gas scrubbers capable of reducing sulfur dioxide emissions by 85 percent. The cost of energy from coal will rise substantially as additional pollution abatement equipment is installed on generating plants and as new environmental regulations lengthen the construction period for coal-fired generating plants. Similar technological and environmental concerns will restrict the development of coal gasification and liquefaction.

Nuclear Fission Power

Roughly 12 percent of the electricity generated in the United States is provided by nuclear power plants.[29] Commitments to plants under construction or on order will bring our nuclear generating capacity to some 200 gigawatts by 1985 and possibly to 300 to 400 gigawatts by 2000. (One gigawatt is 1 billion watts of electric power, roughly the output of a single, large nuclear plant.) Past experience with nuclear power plants has demonstrated that they are capable of generating large quantities of electricity at a lower cost and with lower impact on public health and the environment than any alternative method.[30] Indeed, the present economic and environmental advantages of nuclear power generation seem well established. The real question is whether nuclear power generation will continue to exhibit these advantages in the future, or whether the past experience has been merely fortuitous and the inherent limitations of this technology will become more apparent as it is expanded in the years ahead.

The expansion of nuclear generating capacity has met with a number of serious difficulties. The enormous complexity of federal and state regulations used in licensing nuclear power plants has added greatly to their construction time (typically twelve years) and hence to their cost. A major factor here has been the degree to which the interpretation and application of federal regulations has been decided eventually by the courts through time-consuming litigation. Furthermore the capital-intensive nature of nuclear power generation makes this particular energy source extremely sensitive to the health of the economy since massive amounts of capital must be raised to build nuclear plants. Critics have focused on a number of other drawbacks of nuclear power generation, including low-level radiation releases, nuclear reactor safety, radioactive waste disposal, terrorism and sabotage, and theft of nuclear materials—all issues that threaten to undermine public acceptance of this technology. Events such as the Three Mile Island accident have further eroded public confidence in the safety of nuclear power. Perhaps of most significance, the absence of a clear federal policy on nuclear power development in general and the nuclear fuel cycle in particular has created a climate of uncertainty that has inhibited further expansion of nuclear generating capacity.

Conservation
Regardless of how rapidly we expand our sources of energy, whether by adding new coal-fired or nuclear-powered plants, we will nevertheless need to make a substantial commitment to conservation to balance our energy needs and demands against our energy supply. Exponential growth in energy consumption cannot continue indefinitely.

However, as we noted earlier, our present society is highly dependent on energy. An abrupt transition to a less energy-intensive lifestyle would undoubtedly have a severe impact on our economy. Therefore energy conservation measures must be introduced gradually; they must be stimulated by government policies in such a way as to cause only modest perturbation of existing lifestyles and the economy. Perhaps the most effective method of stimulating conservation is to allow energy prices to rise gradually to levels that more accurately reflect fuel replacement costs. If domestic petroleum prices were allowed to rise to levels more in line with international crude oil prices, then there would be more incentive to improve the efficiency of the transportation industry. It is generally agreed that natural gas prices must be allowed to rise to more realistic levels, thereby forcing the demand for natural gas into line with existing supplies and providing the incentive to increase production of this fossil fuel.

Higher fuel costs would be an economic incentive for the improvement of both supply and end-use technology.[31] Improvement would include increased efficiency of electric generating and transmission processes, lower energy consumption by industry, and more judicious use of energy by commercial and individual consumers. Furthermore higher energy costs would have the positive effect of stimulating development of energy sources and production methods, such as solar heating and oil production from shale, that are now considered economically marginal, thereby increasing energy supply.

The next few years are certain to witness the development and implementation of new technology to achieve greater energy efficiency.[32] Improvements will include lighter weight automobiles, new building designs that take advantage of passive solar heating and improved insulation, new industrial boiler designs for process heat and improved heat recovery processes, and electric load switching and load leveling. These modest improvements will not achieve a dramatic reduction in our energy consumption; rather they will merely reduce the rate of increase in consumption from its present level of 3.5 percent to perhaps 1.5 to 2.0 percent per year.

There have been proposals that our society go one step further and discard the present energy supply and use systems entirely.[33] That is, we should abandon *hard*-energy technologies that rely on centralized electricity generation as the cornerstone of our energy supply and move toward *soft*-energy technologies that are decentralized and rely on benign energy sources such as the sun. The proponents of a soft-energy future assert that the high technology of centralized electricity generation is bankrupt and that its capital-intensive nature, coupled with the escalating costs of conventional fuels, will make it rapidly obsolete. The use of premium fuels and electricity for most uses is wasteful. Only a small fraction of energy end-uses require electricity (about 8 percent at present). Furthermore only a fraction (less than 30 percent) of the energy consumed to produce electricity eventually reaches the consumer because of elaborate energy conversion processes.

To phase over to alternative energy sources, soft-energy proponents propose a dramatic conservation program. They suggest that this country could implement a variety of technical fixes to improve energy efficiency by a factor of 3 to 4 by the year 2000. This program would allow nuclear power to be phased out immediately. We would then rely on large-scale production of energy from coal, oil, and gas until that point in the early twenty-first century when soft-energy technologies such as solar power would be sufficiently advanced to supply all of our energy needs.

A key feature in the argument for a soft-energy future is the assertion that hard- and soft-energy technologies are basically incompatible, that hard- and soft-energy futures are mutually exclusive. Soft-energy proponents assert that the pattern of commitments of resources and time required for the present hard-energy path will gradually make the soft path less and less attainable.

This claim of incompatibility, this refusal to accept any part of a future that contains central station electricity generation (particularly from nuclear power) is the Achilles' heel in soft-energy proposals,[34] for such proposals rely almost entirely on unproven technologies and demand dramatic changes in our lifestyles, changes that almost certainly would not be accepted by our present society. We must keep in mind that, while the conservation of energy resources is certainly important, other factors such as environmental impact, public safety, and economics also are vital factors in the suitability of an energy technology. Furthermore an energy-intensive society has some rather significant advantages. In replacing human labor, energy has provided man with the time to enjoy other pursuits (such as devising and promoting soft-energy futures).[35] Certainly central station electric power generation is inefficient when matched against many end-uses such as space heating, but there is little doubt that it is convenient, reliable, and for the present, relatively inexpensive.

Therefore it is most unlikely that our society will opt for a soft-energy future exclusively and for the low-energy lifestyle that it would entail. We must not pin too many hopes on conservation. Even in the short term, it is highly doubtful that energy conservation can match demand with supply while maintaining our present standard of living, providing for our growing population, and improving the standard of living for the disadvantaged segment of our society.[36] Over the long term conservation alone can, at best, only delay the time when diminishing reserves of conventional energy resources will threaten society and civilization as we know it.

The Short-Term Prognosis
For the short term any realistic energy policy must recognize that there are only three viable options for balancing our energy supply

and demand: coal, nuclear power, and conservation. Because of the basic limitations of each option, all three must be considered essential components of our future energy policy. We must open more coal mines, improve our coal transportation network, and reduce the hazards of coal mining and combustion so we may burn as much coal as we need as rapidly as possible. At the same time we must expand nuclear power generation by developing acceptable technologies for radioactive waste disposal and spent fuel reprocessing, by streamlining regulatory procedures, by improving the methods used to finance nuclear plant construction, and by mounting extensive programs of public education to overcome misunderstandings and fears of this new technology. Finally we must use every possible avenue to stress the importance of energy conservation and provide sufficient economic incentives for massive conservation efforts.

Long-Range Alternatives

Energy sources that are either renewable or characterized by extremely large resource bases are solar power, geothermal power, and advanced forms of nuclear power such as the breeder reactor and controlled thermonuclear fusion. A great many barriers must be overcome before these energy sources can be deployed on a massive scale, that is, before they reach the stage of economic or social viability. Since it seems unlikely that these problems can be solved before the turn of the century, we have classified these options as long range.

Solar Power

In recent years solar power has been promoted as the eventual replacement for both fossil and nuclear fuels in meeting our future energy needs. The conversion of sunlight into useful energy is commonly perceived not only as an exceptionally benign energy source with minimal impact on the environment and public health, but also as an inexhaustible source of inexpensive energy (sunlight is plentiful and free). Solar power has become the cornerstone of proposals to move to a soft-energy society. And even though most forms of solar energy are not yet viable on a massive scale, there seems to be an implicit public faith in the capability of science and technology to develop solar energy resources if sufficient funding is provided.

Certainly the potential of this renewable resource is enormous. The rate at which solar energy falls on the United States is some six hundred times our current consumption rate.[37] To this we can add the large resources of hydroelectric, wind, and ocean thermal energy. However, although solar energy may be plentiful, it will certainly not be cheap.[38] Solar energy is highly diffuse, and it is usually not available at a constant or predictable rate. One must collect, concentrate, and convert solar energy into useful forms, and this will require a rather significant investment. In fact many of the barriers to the massive implementation of solar energy production are not scientific or technological, but rather economic and institutional. Certainly solar energy is scientifically feasible today since such energy systems have been designed, constructed, and operated for a number of years. However, the more serious question is whether solar energy will be commercially viable as a practical component of our nation's energy production capacity in the near future and in competition with alternative means of providing the same energy.

Certain applications of solar energy are rapidly approaching viability. For example, solar heating could be implemented rapidly if there were enough economic incentives to build up a demand and thereby stimulate a manufacturing industry. Although present cost estimates of solar heating systems range from $4,000 to $20,000 per home, mass production methods are being studied that should greatly reduce these costs.[39] However, at the present time, an undeveloped market demand, a technology that is unproven on a commercial scale, and relatively high production costs have inhibited implementation of solar heating.

Of more direct concern is the use of solar energy to generate electricity.[40] Here there are several options, including massive use of photovoltaic cells; solar thermal plants in which huge mirrors focus the sun's rays on boilers that produce steam for conventional electricity generation; huge solar collector satellites placed in orbit that beam their power down to earth; windmill electric generators; use of thermal gradients in ocean currents to power electric plants; and biomass conversion of large quantities of vegetation into liquid or gaseous fuels by chemical processes. Unfortunately these schemes to produce electricity from solar energy must surmount many hurdles before they can be deployed on a massive scale.

The major barriers to the implementation of solar energy arise from its dilute nature.[41] Highly capital-intensive systems utilizing enormous quantities of materials will be required to capture and convert the sun's energy into useful forms. Since these systems must have operating lifetimes of twenty to thirty years to pay off construction costs, their development is a significant technological as well as economic challenge.

Many solar energy systems now under development may not

even be able to regenerate the energy required for their construction.[42] Solar energy, whether for heating or electric power generation, consumes significant quantities of nonrenewable resources in the production of materials for its generation and storage. The time required for the solar energy system to pay back this energy investment may be a significant fraction of its operating life. For example, the present energy payback times for solar heating collectors is about ten years, for solar thermal plants in the desert southwest about five years, and for solar photoelectric plants about fifteen to twenty years, compared with a payback period of one to three years for coal or nuclear power plants. For solar energy to become viable, we must develop the manufacturing capability to produce the large quantities of materials required by solar systems in a far more economic and energy-efficient manner.

Solar energy systems will never be stand-alone systems because of the intermittent nature of sunlight as an energy source. There will always be a need for backup systems, be they fossil-fuel or nuclear, as well as storage devices, and this will significantly increase the capital costs of solar energy systems.

Recent studies also have suggested that solar energy may not be an entirely benign technology.[43] True, the risks from the solar energy system itself may be low, although routine accidents such as falling off the roof while cleaning snow off solar collectors will take their toll. However, when all parts of the energy cycle are compared on a unit of energy production basis, the risks of solar power appear to be comparable to, if not somewhat greater than, those of more conventional energy sources such as nuclear power. These risks arise from the manufacturing processes required to produce the large quantities of materials required for solar systems and from the construction of these massive or numerous systems. The tasks of mining the coal, iron, and other raw materials and fabricating them into steel, copper, and glass give rise to significant public risks. For example, the production of the metals needed in solar energy systems requires energy that will be provided, for the most part, by coal-fired generating plants. This coal will produce air pollution, that in turn causes public health effects. Additional risks will be contributed by the necessary backup and storage systems. And the environmental impact of solar energy systems will certainly not be negligible. One need only imagine a solar electric station covering 10 to 20 square miles with solar collectors or hundreds of windmills dotting the landscape.

Certainly the sun represents an abundant source of energy that may someday provide a significant fraction of mankind's energy needs. For this reason the technology involved in exploiting solar energy should be developed most vigorously. However, the successful development of solar energy as a viable alternative will require the solution of some rather significant technical, economic, and institutional problems. Clearly solar energy technologies have a long way to go before even a small fraction of their enormous potential will be realized.[44]

Geothermal Energy

Geothermal energy is the thermal energy contained in the upper 10 kilometers of the earth's crust.[45] Unfortunately most heat from the earth's interior is too diffuse to be exploited on a wide basis. Hence geothermal resources suitable for commercial purposes are usually regarded as localized deposits of heat concentrated at accessible depths, at confined volumes, and at sufficient temperatures for its intended use. To date the most highly developed geothermal resources are natural sources of high-temperature dry steam such as that used in the Geysers plant in northern California. However, some consideration has been given to using wet-steam sources, or tapping large volumes of geofluids trapped under high pressure and temperature, or circulating water through dry hot-rock formations.

But suitable geothermal sites are limited in number and location. Most experts feel that the potential of this energy source is roughly comparable to that of hydroelectric power generation.[46] Because of the low-grade nature of the heat produced by geothermal deposits, the thermal energy extracted cannot be transported far and must be used directly at the reservoir site or converted into a transportable form such as electricity. The low temperature of such sources leads to a rather low efficiency and hence large quantities of waste heat. Geothermal energy plants exhibit enormous local environmental impact[47] since they release significant quantities of noxious gases such as hydrogen sulfide, liquids, and solids. They cause substantial land settlement and seismic acitivity. A geothermal reservoir is quite limited in capacity, and the output from geothermal sites decreases rather significantly after ten to twenty years of exploitation.

Therefore, although geothermal sources will certainly contribute to our electric generating capacity during the next few decades, they cannot be expected to have a major impact. Geothermal energy should not be regarded as having the same potential as other options such as solar power or nuclear fusion.

Advanced Forms of Nuclear Power

Estimates suggest that domestic uranium resources[48] are sufficient to support the present generation of nuclear power plants based on light water reactor technology until shortly after the turn of the century, assuming an expansion to 400 to 500 gigawatts of nuclear capacity by that time. Hence present nuclear power technology is limited by its fuel resource base to a short-term energy source.

If nuclear power is to play a significant role in the twenty-first century, it will be essential to develop advanced nuclear reactor types known as breeder reactors.[49] Such systems are designed to make far more efficient use of uranium and thorium fuels by converting these into new elements such as plutonium that can be used more directly as nuclear fuels. Indeed, the implementation of breeder reactors would expand the resource base for nuclear power by a factor of almost 50; moreover it would permit the exploitation of even low-grade ore deposits on an economic basis. Therefore the introduction of breeder reactors would extend the viability of nuclear power from perhaps fifty years to thousands of years.

Breeder reactors have been a feasible technology for almost thirty years. In fact the first electric power produced from nuclear fission was provided by a fast breeder reactor (the Experimental Breeder Reactor I in Idaho) in 1951. A large number of experimental breeder reactors have been constructed and operated since then throughout the world. Most nations with a significant commitment to nuclear power (for example, France, West Germany, the U.K., the USSR, Japan) are moving rapidly to develop commercially viable breeder reactors to be introduced within the next decade.

However, breeder reactor technology is plagued by political problems. Since breeder reactors produce and utilize large quantities of fuels (plutonium) that, in principle at least, could be fabricated into nuclear weapons, there is a serious concern that the spread of breeder reactor technology will accelerate the proliferation of nuclear weapons.[50] This concern has significantly slowed the pace of breeder reactor development in this country, and we will examine this important issue in more detail in chapters 4 and 5. However, we should note here that the breeder reactor is the only energy technology capable of long-term implementation (being characterized by essentially infinite fuel resources) that can be expected to reach economic viability within the next decade.

A considerably more exotic, yet far more complex, approach to nuclear power is controlled thermonuclear fusion.[51] This scheme involves fusing together the nuclei of hydrogen atoms at enormous temperatures in such a way that energy is produced. Thermonuclear fusion is frequently promoted as the ultimate answer to mankind's energy needs for all time to come. Theoretically at least, it is characterized by an infinite fuel supply (it would burn the hydrogen in the oceans), and its proponents suggest that it should have marked advantages in safety and environmental impact over nuclear fission. However, thermonuclear fusion is not the panacea for society's energy ills that it might first appear to be.

The major drawback to nuclear fusion is that it doesn't work yet. We have yet to demonstrate the scientific feasibility of this scheme, and we will probably not be able to do so for several years. Beyond that demonstration, it will take a minimum of several decades of engineering research and development to bring fusion power to an economically viable state. In many ways fusion will exhibit safety and environmental effects similar to those of conventional nuclear fission.[52] For example, fusion plants will have a rather large inventory of radioactive material. They will depend on thermal cycles, with their intrinsic inefficiency, to generate electricity. The plants will release small amounts of radiation during normal operation and will produce radioactive waste that must be disposed of in some suitable manner. Despite these similarities, fusion power systems do appear to have a potential for reducing these impacts on man and his environment far below those of fission plants.

Looking ahead to large fusion power reactors, one is almost overwhelmed by their apparent complexity. Fusion reactors are expected to represent an increase in complexity over fission reactors that is roughly comparable to the increased complexity of reactors over coal-fired plants.[53] Certainly the potential of nuclear fusion power is great, but even its most optimistic proponents do not expect it to reach a commercially viable stage until early in the next century.

The Long-Term Prognosis
This rather pessimistic picture of long-range alternatives is not intended to discourage the reader, but rather to place in perspective the enormous problems that we must face and overcome in developing any new energy technology. These problems must be faced realistically and taken into account in the formulation or assessment of energy policy. Massive deployment of future energy technology will require an enormous commitment of intellect, manpower, and money—and even then it will require several decades and a good deal of luck. The extent of this commitment should not be taken lightly. One cannot magically bring solar power or nuclear fusion to the stage of social viability by simply pouring a large quantity of

money and manpower into its development. Rather one must regard these options as *possible* long-range energy alternatives and make plans for the near term accordingly. It would be foolhardy to rely on false, or at best unsubstantiated, hopes for these long-range technologies and postpone the massive implementation of the presently viable energy sources—coal and nuclear power—to meet short-term energy needs.

The Need for Nuclear Power

Let us now return to the question we posed at the beginning of this chapter: Is there really a need for nuclear power? We have briefly examined the so-called energy crisis, the growing imbalance between our demand for energy and our resources for providing this energy. It is apparent that our national appetite for energy cannot continue to grow at its historical rate. There is simply no way we can continue to satisfy this exponentially increasing energy consumption indefinitely. Rather it is obvious that this nation must limit its growth in energy demand.

At the same time it is highly unlikely that our energy consumption will decline or even level off in the near future in the face of a growing population and a desire for upward mobility by those disadvantaged members of our society. When growing demand is coupled with a decline in domestic and eventually worldwide production of natural gas and petroleum that seems likely to occur by the end of the next decade, it seems apparent that we should move rapidly to develop and implement alternative energy sources.

For the short term our options are quite limited. Coal combustion and nuclear power are presently the only viable technologies capable of massive implementation before the end of this century. Other energy technologies, such as solar, wind, biomass, geothermal, and nuclear fusion power, are essentially unproven, although in some cases feasible, energy sources. Certainly the *potentials* of these alternative energy technologies are outstanding, and the research and development necessary to bring them to a viable stage should be pursued with all vigor. But we would be most foolish to depend solely on the potential of these options to meet future, even long-range, energy needs. The future is fraught with far too many uncertainties for us to narrow our options now by discarding a technology such as nuclear power that can meet a significant fraction of our energy needs.

Hence the real choice is not whether to use nuclear power, but rather the balance between our dependence on nuclear power and coal to meet our future demand for electricity. Both sources will be required, regardless of our success in implementing energy conservation measures.

But what about the potential hazards of nuclear power—the dangers of a nuclear plant accident resulting in large numbers of public casualties, or the disposal of radioactive waste, or the international implications of spreading a nuclear technology that can be used for either war or peace? Certainly these perceived hazards are serious—but they are also hypothetical. In sharp contrast the hazards from mining and burning coal are very real. The great quantity of coal combustion necessary to meet a significant fraction of our energy demand will involve a major environmental impact and public risk. There is absolutely no doubt about this.

While the perceived dangers of nuclear power should certainly be considered, they are quite different from those characterizing other energy options. In a sense we are balancing "what if" against "what is" concerns, potential versus actual hazards. Furthermore nuclear power technology has already demonstrated during two decades of operating experience that it is possible to design nuclear power systems to minimize these risks, to keep them hypothetical, and to protect the public from possible consequences of even serious accidents (e.g., Three Mile Island).

In future chapters we will examine those aspects of nuclear power generation that are commonly perceived as the primary drawbacks of this technology. However, we should make it clear at this point that we accept the implementation of nuclear power technology as necessary if we are to maintain our society and lifestyle as we now know them. We regard nuclear power generation as a proven technology, characterized by significant advantages over alternative energy technologies in minimizing public risks, environmental impact, and costs of generating electricity. To discard nuclear power entirely would seem to us to be a most imprudent and unfortunate action.

2

The Development of Nuclear Power

For the past three decades an extensive international effort has been directed toward harnessing the enormous energy within the atomic nucleus for the peaceful generation of electric power. Nuclear reactors have evolved from research tools into mammoth units that drive hundreds of central station electric generating plants throughout the world today. The impending shortage of fossil fuels has cast even more significance on the role that nuclear power must play in meeting man's future energy requirements.

The size and cost of a modern nuclear power plant such as that shown in figure 3 are truly staggering. Such plants typically generate over 1 billion watts of electric power, an amount sufficient to supply the needs of a city of almost a million people. The plants cost more than $1 billion to construct. The Department of Energy estimates that, if present trends continue, some 380 nuclear power plants will be installed in the United States by the year 2000, representing a capital investment of more than $500 billion.[1] Industrialized nations throughout the world are making commitments to nuclear power technology at an even more rapid pace.

In this chapter our discussion will range from fundamental concepts of nuclear fission chain reactions to descriptions of principal types of nuclear power systems used today. We will trace the historical development of nuclear power from its early days as the stepchild of the top-secret Manhattan Project to the maturation of today's nuclear power industry. We will also examine the growth in public opposition to this new energy source and briefly catalog the various issues that have arisen in the debate over nuclear power.

Basic Concepts of Nuclear Power Generation

The term *nuclear reactor* refers to devices in which controlled nuclear fission chain reactions can be maintained. In such devices

Fig. 3. The Enrico Fermi Nuclear Power Station, Unit Number 2, now under construction south of Detroit, Michigan. (*Courtesy of the Detroit Edison Company.*)

neutrons are used to induce nuclear fission reactions in the atomic nuclei of heavy metals such as uranium or plutonium. These nuclei fission into lighter nuclei, which are known as fission products. The fission process is accompanied by the release of energy plus several additional neutrons. The neutrons produced in the fission reaction can then be used to induce further fission reactions, thereby propagating a chain of fission events. This process is shown in figure 4. In a very narrow sense, a nuclear reactor is simply a sufficiently large mass of appropriately fissile material, such as the isotopes* uranium-235 or plutonium-239, in which a controlled *fission chain reaction* can be sustained. A sphere of U-235 metal slightly over 8 centimeters in radius could support such a chain reaction and therefore would be classified as a nuclear reactor.

---

* An *isotope* is a form of an element whose atomic nuclei are characterized

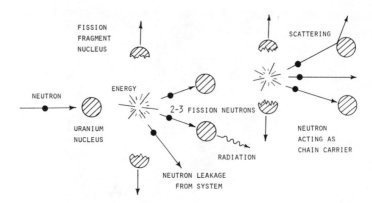

FISSION
FRAGMENT
NUCLEUS

SCATTERING

NEUTRON

ENERGY

2-3 FISSION NEUTRONS

URANIUM
NUCLEUS

NEUTRON
ACTING AS
CHAIN CARRIER

RADIATION

NEUTRON LEAKAGE
FROM SYSTEM

Fig. 4. A fission chain reaction.

But a modern power reactor is considerably more complex than a simple sphere of metal. It contains not only a lattice of carefully refined and fabricated nuclear fuel, but also the apparatus for cooling this fuel during the course of the chain reaction as fission energy is released while maintaining the fuel in a precise geometric arrangement with appropriate structural materials. Furthermore some mechanism must be provided to control the chain reaction. The surroundings of the reactor must be shielded from the intense nuclear radiation generated during the fission reactions. Fuel handling equipment is necessary for loading in and replacing nuclear fuel assemblies when the fission chain reaction depletes the concentration of fissile nuclei. If the reactor is to produce power in a useful fashion, it must also be designed so that it will operate economically, reliably, and safely. These engineering requirements make the actual configuration of a nuclear power reactor quite complex indeed.

Fission Chain Reactions and Nuclear Criticality
To understand the principal concepts underlying nuclear reactor operation, we need to look at the fission chain reaction process.[2] To maintain a stable fission chain reaction, a nuclear reactor must be designed so that, on the average, exactly one neutron from each fission will induce yet another fission reaction. That is, the production of neutrons from fission reactions must be balanced against their

by a particular number of protons and neutrons, referred to collectively as nucleons. For example, uranium occurs in several isotopic forms possessing 233, 234, 235, 236, or 238 nucleons. It is convenient to abbreviate the notation for a particular isotope by using the chemical symbol for the element followed by the number of nucleons. For example, uranium-235 is written U-235.

loss by either leakage from the reactor or absorption in nuclear reactions that do not lead to fission.

For example, suppose that in a particular nuclear system, more neutrons are lost by leakage and absorption than are produced in fission. A self-sustaining chain reaction cannot be achieved, and we say the system is *subcritical*. One way to alter the system so that there is a more favorable balance between production and loss is simply to make it bigger. Then the probability that a neutron will leak out before being absorbed by a nucleus is decreased, since the average distance a neutron has to travel to leak out increases in such a way that the neutron will undergo more collisions on the way. An alternative approach would be to increase the relative concentration of fissile nuclei. By adjusting the fuel concentration and the size of the reactor, we can balance neutron production versus loss and achieve a self-sustaining chain reaction, a *critical* system. Figure 5 illustrates this process.

Actually it is appropriate to dismiss neutron leakage and reactor size from further consideration since most modern power reactors are so large that few neutrons leak out, usually less than 3 percent. In fact the size of a power reactor is determined not by the desire to minimize neutron leakage, but rather to provide enough space for coolant flow to remove adequately the enormous heat produced by the fission reactions. In reactor design one first determines how large the reactor must be to accommodate adequate cooling for a desired power output. Then one determines the fuel concentration that will yield a critical reactor of this size.

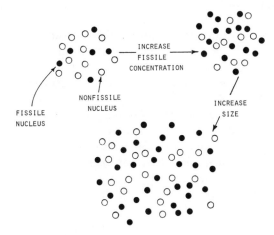

Fig. 5. Achieving a critical mass by increasing fissile concentration and system size.

Suitable Fuels for Fission Chain Reactors
The necessary fuel concentration will depend sensitively on the fuel
type. Nuclear engineers characterize the suitability of a material for
sustaining a fission chain reaction by a parameter denoted by the
Greek symbol eta.

> η = average number of neutrons produced by fission
> per neutron absorbed by fuel nucleus.

This parameter not only characterizes the relative propensity of a
fuel nucleus to fission, but also its ability to shed further neutrons in
the fission process that can be used to sustain the chain reaction.
Evidently, if the fission chain reaction is to proceed in a self-
sustained manner, we must utilize fuels that are characterized by
values of η greater than 1 since some neutrons will always leak out
or will be absorbed in nonfuel materials in the reactor.

There are few heavy nuclei that have values of η sufficiently
greater than 1 to be of interest as candidates for nuclear reactor fuel.
These *fissile* materials include the isotopes of uranium, U-233 and
U-235, and plutonium, Pu-239 and Pu-241. Unfortunately only the
isotope U-235 occurs in nature, and then only as a small percentage
(0.711 percent) of naturally occurring uranium, which is primarily
composed of U-238. To utilize uranium as a reactor fuel effectively,
it is usually necessary to increase the concentration of U-235, that is,
to "enrich" natural uranium in this isotope through the use of
elaborate and expensive isotope separation methods.

The other fissile isotopes can be produced artificially by bombarding certain materials with neutrons. For example, uranium-238
and thorium-232 can be transmuted into the fissile isotopes Pu-239
and U-233 respectively by exposing them to neutrons in a reactor. In
fact there are usually a sufficient number of excess neutrons produced in a fission chain reaction (since η is greater than 1) so that
fertile materials such as U-238 can be transmuted into fissile isotopes
such as Pu-239 as the reactor operates. Actually this conversion of
fertile into fissile material occurs in all modern power reactors since
they contain substantial amounts of U-238, which will be transmuted
into plutonium during normal operation. For example, a light water
reactor will contain a fuel mixture of roughly 3 percent U-235 and 97
percent U-238 in freshly loaded fuel assemblies. After a standard
operating cycle of three years, this fuel will contain roughly 1 percent U-235 and 1 percent plutonium, which could then be separated
out of the spent fuel and refabricated into fresh fuel elements for
reloading. This latter process is referred to as *plutonium recycle*.

These considerations suggest that it might be possible to fuel a reactor with Pu-239 and U-238 and then produce directly the fuel (Pu-239) needed for future operation. Indeed, it is even possible to produce more Pu-239 than is burned, that is, to "breed" new fuel, if η is large enough. To be more precise, one can introduce the *conversion ratio*:

$$CR = \frac{\text{average rate of fissile nuclei production}}{\text{average rate of fissile nuclei consumption}}$$

From this definition it is apparent that consuming N atoms of fuel during reactor operation will yield $CR \times N$ atoms of new fissile isotopes. Most modern light water reactors are characterized by a conversion ratio of about 0.6. In contrast, gas-cooled reactors have a somewhat higher conversion ratio of 0.8 and are sometimes referred to as *advanced converter* reactors. For breeding to occur, the conversion ratio must be greater than 1—in which case we rename it the *breeding ratio*. For this to happen η must be greater than 2, since slightly more than one fission neutron is needed to maintain the chain reaction (if we account for neutron leakage or parasitic capture), while one neutron will be needed to replace the consumed fissile nucleus by converting a fertile into a fissile nucleus.

To achieve this we take advantage of the fact that the parameter η depends not only on fuel type, but on the average speed or kinetic energy of the neutrons sustaining the chain reaction as well. In general, η becomes larger as the neutron energy increases. Since the average energy or speed of fission neutrons is quite large (*fast* neutrons), one can maximize the value of η and hence the breeding ratio by simply designing the reactor so that these neutrons do not slow down during the chain reaction. Reactors that optimize the breeding of new fuel by utilizing fast neutrons are known, naturally enough, as *fast breeder reactors*. A more careful comparison of η for various fissile isotopes[3] makes it apparent that the optimum breeding cycle for fast reactors would utilize U-238 as the fertile material and plutonium as the fissile fuel. Careful breeder reactor designs can achieve breeding ratios of 1.3 to 1.5 based on this cycle.

But there is a significant drawback to such reactor designs. The probability of a neutron inducing a fission reaction decreases with increasing energy, so that the minimum fuel is required for a chain reaction sustained with low energy or *slow* neutrons. For this reason most power reactors are designed so that the fast fission neutrons are slowed down or *moderated* to enhance the probability that they will induce fission reactions. In such reactors materials of low mass number such as water or graphite are interspersed among the fuel elements. Then, as the fission neutrons collide with the nuclei of these moderator materials, they rapidly slow down to energies comparable to the thermal energies of the nuclei in the reactor core—which explains the term *thermal reactors* used to describe such designs. Most nuclear power plants utilize thermal reactors since these systems require the minimum amount of fissile material for fueling and are the simplest reactor types to build and operate, even though they are incapable of achieving a net breeding gain.

It is possible, at least in theory, to achieve breeding even in thermal reactors, if a thorium-232/uranium-233 fuel cycle is used.[4] Unfortunately the maximum breeding ratio one can achieve in practice with this cycle appears to be only slightly greater than 1 (CR is about 1.02). Although this marginal breeding ratio would lead to a more effective utilization of nuclear fuel, at least in comparison with conventional thermal reactor types such as light water reactors, it would not produce enough additional fissile material (U-233) to fuel new reactors. A light water breeder reactor based on the thorium/U-233 fuel cycle went into operation at Shippingport, Pennsylvania, in 1977.

Nuclear Reactor Operation

The final parameter of interest in fission reactor design is the fuel concentration, which can be chosen to balance neutron production (fission) and loss (leakage and absorption). The fuel concentration can be altered by adjusting the *enrichment* of the fuel (that is, the percentage of fissile material in the fuel), the density and geometry of the fuel, and the quantity of nonfuel materials in the core. One refers to the amount of fissile material required to achieve a critical fission chain reaction as the *critical mass* of the fuel. This amount depends sensitively on the particular composition and geometry of the fuel. For example, the critical mass of a sphere of pure U-235 metal surrounded by a natural uranium reflector is only 17 kilograms. In contrast, the fuel loading required for a modern light water reactor is roughly 100 tons of 3 percent enriched uranium.

It should be noted, however, that a nuclear power reactor is always loaded with much more fuel than is required merely to achieve criticality. The extra fuel is included to compensate for those fuel nuclei destroyed in fission reactions during power production. Since most modern reactors are run roughly one year between refuelings, a sizable amount of excess fuel is needed to compensate for fuel burnup as well as to facilitate changes in the reactor power level.

To compensate for this excess fuel, that is, to readjust the balance between neutron production and loss, one introduces materials into the reactor that absorb neutrons from the chain reaction, thereby cancelling out the excess fuel. These absorbing materials can then slowly be withdrawn as the fuel burns up. They can also be used to adjust the criticality of the nuclear reactor in such a way as to control the chain reaction. This might be accomplished in a variety of ways. For example, the neutron absorber might be fabricated into rods that can then be inserted into or withdrawn from the reactor at will to regulate the power level of the reactor. That is, one can withdraw the absorber or *control rods* to make the reactor slightly supercritical so that the chain reaction builds up. Then when power has reached the desired level, the rods can be reinserted to achieve a critical or steady-state chain reaction. Finally the rods can be inserted still further into the reactor to shut the chain reaction down.

The longer term changes in fuel concentration due to burnup are usually compensated by neutron absorbers fabricated directly into the fuel or dissolved in the coolant. These absorbers or poisons will burn up as the fuel burns up, thereby balancing out the excess core fuel. Eventually the fuel will be so depleted that the reactor can no longer be made critical, even by withdrawing all of the control rods, and the reactor must then be shut down and reloaded with fresh fuel.

One important facet of reactor operation concerns the stability of the reactor. The probabilities for various neutron interactions (fission or absorption) depend quite sensitively on the temperature and therefore on the power level of the reactor. Nuclear reactors are designed with negative feedback so that increasing power, and therefore temperature, brakes the chain reaction, thereby decreasing power again. The most significant feedback mechanisms involve a decrease in moderator density, and therefore a decrease in moderation, with increasing temperature in thermal reactors and an enhanced tendency of nonfissile material to absorb neutrons with increasing temperature in all reactor types. These mechanisms are so strong that reactors tend to operate in a stable fashion. In fact most power reactors are quite incapable of operating significantly above their designed power level. That is, if somehow the reactor was inadvertently made supercritical—a control rod was accidently withdrawn while the reactor was operating at full power, for example—the power level of the reactor and hence the reactor temperature would increase only slightly before negative feedback would return the reactor to a critical or subcritical state by slowing down the chain reaction. Such inherent feedback mechanisms, coupled with the dilute nature of the reactor fuel, eliminate any possibility of a runaway chain reaction and remove the concern about possible nuclear accidents involving the chain reaction itself from nuclear power reactor design and operation.

Although a nuclear reactor cannot explode like a bomb, it

generates enormous quantities of potentially hazardous radioactive material (primarily fission products). If somehow the reactor were to crack open and release this radioactivity into the environment, it could pose a significant danger to public health. Therefore a number of physical barriers are designed into the reactor to contain radioactive materials and prevent their release to the environment under any conceivable condition. Precautionary measures include cladding the fuel pellets in metal tubes, encasing the reactor core in a steel pressure vessel, and surrounding it all with concrete shielding and steel-lined, reinforced concrete walls.

The most serious reactor accident would then be some sequence of events that might breach this containment and release radioactivity to the environment. Typically this is imagined to be a core meltdown accident in which the reactor core suddenly loses cooling, due to a pipe break, for example. Although the chain reaction would shut down immediately because of the loss of moderation, the heat produced by the radioactive decay of fission products would be sufficient to melt the reactor fuel elements, if auxiliary cooling were not provided. The molten fuel could then slump to the bottom of the reactor vessel, melting through it and the concrete floor of the reactor building, and releasing fission product gases to the environment. In chapter 3 we will discuss how so-called engineered safety systems are incorporated into reactor design to prevent such an occurrence.

Nuclear Power Generation
Nuclear power reactors are designed to produce heat that can then be used to generate electric energy, usually by way of an associated steam thermal cycle. In this sense the primary function of a reactor is that of an exotic heat source for turning water into steam. Aside from the nuclear reactor and its associated coolant system, nuclear power plants are remarkably similar to large fossil-fuel-fired power plants (see fig. 6). Only the source of the heat energy differs, nuclear fission versus chemical combustion. Most components of large central station power plants are common to both nuclear and fossil-fueled units.

The current generation of power plants operates on a steam cycle in which the heat generated by combustion or fission is used to produce high-temperature steam. This steam is then allowed to expand against the blades of a turbine. In this way the thermal energy of the steam is converted into the mechanical work of turning the turbine shaft. This shaft is connected to a large electric generator that converts the mechanical turbine energy into electric energy that can be distributed over an electric power grid. The low-pressure

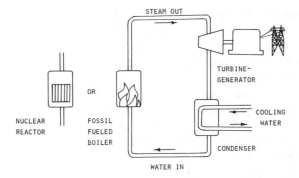

STEAM OUT

TURBINE-
GENERATOR

NUCLEAR
REACTOR

OR

FOSSIL
FUELED
BOILER

COOLING
WATER

CONDENSER

WATER IN

Fig. 6. A simple schematic of a steam-driven electric power plant, either nuclear- or fossil-fueled.

steam leaving the turbine must be recondensed into liquid in a steam condenser so that it can be pumped back to the steam supply system to complete the cycle. The condenser requires large quantities of cooling water at ambient temperature, which are usually obtained from artificial cooling ponds or cooling towers.

Nuclear Steam Supply Systems

At the heart of a nuclear power plant is the *nuclear steam supply system*, which produces the steam used to run the turbine generator. The system consists of three major components: (1) the *nuclear reactor* that supplies the fission heat energy, (2) several *primary coolant loops* and *pumps* that circulate a coolant through the nuclear reactor to extract the fission heat, and (3) heat exchangers or *steam generators* that use the heated primary coolant to turn water into steam. The nuclear steam supply system in a modern nuclear power plant is completely enclosed within a containment structure designed to prevent the release of radioactivity to the environment in the event of a gross failure of the primary coolant system. This nuclear island within the plant is the analog to the boilers in a fossil-fueled unit.

The primary component of the steam supply system is the nuclear reactor. Far from being just a simple pile of fuel and moderator, a modern power reactor is an enormously complicated system designed to operate under the severest conditions of temperature, pressure, and radiation. The energy released by nuclear fission reactions appears primarily as kinetic energy of the various fission fragment nuclei. The bulk of this fission product energy is rapidly deposited as heat in the fuel. This heat is then extracted by a primary coolant flowing between the fuel elements and transported (convected) by this coolant to the steam generators.

A variety of coolants can be used in the primary loops of the steam system. In fact nuclear reactor types are usually characterized by the type of coolant they use, such as "light water" reactors or "gas-cooled" reactors. There are also various possible steam system configurations. For example, one may actually produce the steam in the reactor itself. Or one may use a liquid or gas primary coolant such as high-pressure water or helium to transfer the fission heat energy to a steam generator. In the liquid-metal-cooled fast breeder reactor, an intermediate coolant loop must be used to isolate the steam loop from the high induced radioactivity of the primary coolant loop that passes through the reactor.

The most common coolant used in power reactors today is ordinary water, which serves as both a coolant and a moderating material. There are two major types of *light water reactors* (LWR):[5] *pressurized water reactors* (PWR) and *boiling water reactors* (BWR). In a pressurized water reactor (see fig. 7) the primary coolant is water, maintained under high pressure (150 times atmospheric pressure) to allow high coolant temperatures (300°C) without steam formation within the reactor. The heat transported out of the reactor core by the primary coolant is then transferred to a secondary loop containing the working fluid (the steam system) by a heat exchanger known as a steam generator, since it is within this component that the inlet water is converted into steam. Such systems typically contain from two to four primary coolant loops and asso-

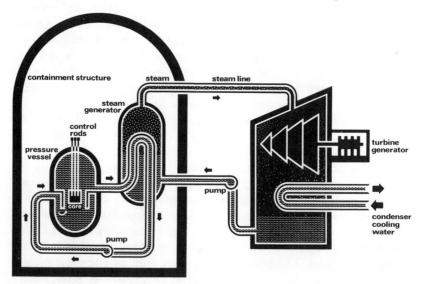

Fig. 7. A pressurized water reactor. (*Courtesy of the Atomic Industrial Forum.*)

ciated steam generators. In addition a surge chamber or pressurizer in the primary coolant loop maintains the high primary system pressure as it accommodates coolant volume changes in the primary loop. The primary loop also contains coolant pumps, as well as auxiliary systems that control coolant purity and inject new makeup water or control absorbers into the coolant.

In a boiling water reactor (see fig. 8) the primary coolant water not only serves as moderator and coolant, but also as working fluid since the system pressure is kept sufficiently low (70 atmospheres) that appreciable boiling and steam formation can occur within the reactor. In this sense the reactor itself serves as the steam generator, thereby eliminating the need for a secondary loop and heat exchanger. Since there is an appreciable steam volume in the primary loop, a pressurizer tank is not required to accommodate pressure surges.

The coolant water rising to the top of the BWR core is a wet mixture of liquid and vapor. Therefore moisture or steam separators must be used to separate off the steam, which is then piped outside the reactor pressure vessel to the turbine and then through the condenser, before it is pumped back into the core as liquid condensate. The saturated liquid that is separated off by the moisture separators flows downward around the core and mixes with the return condensate. This natural recirculation is assisted by pumps.

In both PWR and BWRs the nuclear reactor itself and the

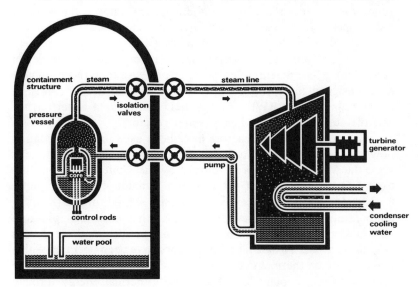

Fig. 8. A boiling water reactor. (*Courtesy of the Atomic Industrial Forum.*)

primary coolant are contained in a large steel pressure vessel designed to accommodate the high coolant pressures and temperatures. In a PWR this pressure vessel must be fabricated with thick steel walls to contain the high primary coolant pressures. The BWR pressure vessel need not be so thick-walled, but it must be larger to contain both the nuclear reactor and steam moisture separating equipment.

The direct cycle involved in a BWR does present one major disadvantage. Since the working fluid actually passes through the reactor core before passing out of the containment structure and through the turbine, one must be particularly careful to avoid radiation hazards. The primary coolant water must be carefully treated to remove any impurities that might become radioactive when exposed to the neutrons in the reactor. Even with this purification, the primary coolant will exhibit significant induced radioactivity, and therefore the turbine building must be heavily shielded.

A closely related class of reactors uses heavy water, D_2O, as moderator and either D_2O or H_2O as primary coolant.[6] The most popular heavy water reactor is the CANDU-PHW reactor (Canadian deuterium uranium pressurized heavy water reactor). This reactor uses a pressure tube design in which each coolant channel in the reactor accommodates the primary system pressure individually, thus eliminating the need for a pressure vessel. As with a PWR the primary coolant thermal energy is transferred by means of a steam generator to a secondary loop containing light water as the working fluid. One major advantage exhibited by heavy water reactors is their ability to utilize natural uranium (with only 0.711 percent U-235) as fuel due to the superior neutron moderating properties of deuterium. More recently heavy water pressure tube reactors have been designed that produce H_2O steam directly in the core in a manner similar to a BWR (the CANDU boiling light water reactor and the steam-generating heavy water reactor).

Gas-cooled nuclear reactors have been used for central station power generation for many years. The earliest such power plants were the Magnox reactors developed by the United Kingdom that used CO_2 as the coolant for a natural uranium-fueled, graphite-moderated core. More recently interest has shifted toward the high-temperature gas-cooled reactor (HTGR) that uses high pressure helium to cool an enriched uranium/thorium core moderated with graphite (fig. 9).[7] To date all such reactors have been operated with a two-loop steam thermal cycle similar to that of a PWR in which the primary helium coolant loop transfers its thermal energy through steam generators to a secondary loop containing water as the working fluid. The HTGR is capable of operating at relatively high temperature, thereby producing high-temperature (400°C), high-pressure (60 atmospheres) steam with an attendant increase in thermodynamic efficiency. Moreover HTGRs have the potential for running in a direct cycle configuration using high-temperature helium to drive a gas turbine (with thermal efficiencies approaching 50 percent).

The HTGR exhibits several other advantages. The use of helium as a coolant not only allows higher operating temperatures at moderate pressures, but also provides flexibility in the selection of optimum coolant temperature, pressure, and flow rate conditions. It also mitigates effects of any loss of coolant accident. Since the coolant always remains in the same phase, the worst that can happen in the event of a rupture of the primary coolant loop is a loss of pressure. And in an HTGR only a modest circulation of helium at atmospheric pressure is needed to remove the radioactive decay heat given off by the core following shutdown. The higher operating temperatures in the HTGR give an edge to a U-235/Th-232/U-233 fuel cycle over the enriched uranium/plutonium fuels of light water or heavy water reactors (although these latter reactor types could also be fueled with a uranium/thorium mixture).

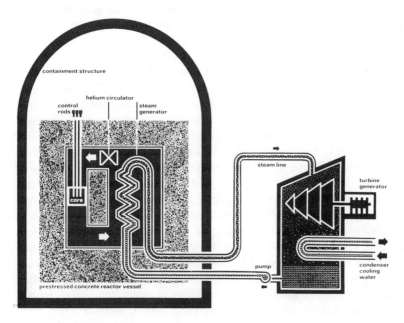

Fig. 9. A high-temperature gas-cooled reactor. (*Courtesy of the Atomic Industrial Forum.*)

The gas coolant does lead to low power densities and therefore large reactor sizes. Furthermore, since the fissile material in such reactors is highly enriched U-235 (which is then mixed with thorium), the HTGR presents a rather major problem from the viewpoint of proliferation of nuclear weapons materials. Nevertheless these reactors have been under development in both the United States and West Germany.

Gas coolants have also been proposed for use in fast breeder reactors (the gas-cooled fast reactor or GCFR). Because of the very high power density required by such reactors, extremely high coolant flow rates would be required. Nevertheless the large breeding ratios (CR = 1.5) achievable in the GCFR make it a promising alternative to other fast reactor designs that utilize liquid metals such as sodium as primary coolant.

Although sodium could be used in thermal reactors if alternative moderation were provided, its primary advantages occur in fast breeder reactors that require a primary coolant with low moderating properties and excellent heat transfer properties.[8] We have noted that the nuclear steam supply system for the liquid-metal-cooled fast breeder reactor (LMFBR) actually is a three-loop system since an intermediate sodium loop must be used to separate the highly radioactive sodium in the primary loop from the steam generators (fig. 10). We will consider this reactor type further in chapter 6.

Nuclear Reactor Components

To introduce the components and systems that make up a nuclear power reactor, we will consider a large, modern boiling water reactor. The reactor proper consists of a *core* containing the fuel, coolant channels, structural components, control elements, and instrumentation systems. The core is a cylinder-shaped lattice roughly 350 centimeters in height consisting of long *fuel assemblies* or *bundles*. As shown in figure 11, each fuel assembly is composed of several hundred long metal tubes, the *fuel elements*, that contain small ceramic pellets of uranium dioxide. Most modern power reactors use such ceramic fuels as either an oxide, carbide, or nitride to facilitate high-temperature operation. The fuel element tube or *cladding* is either stainless steel or a zirconium alloy, which not only provides structural support for the fuel, but also retains any radioactivity produced in the fuel during operation. The primary coolant then flows up through the fuel assemblies between the fuel elements, extracting fission heat. Fuel is loaded into a reactor core or is replaced one fuel assembly at a time. A typical power reactor core will contain hundreds of such fuel assemblies.

The reactor core itself, the structures that support the core fuel

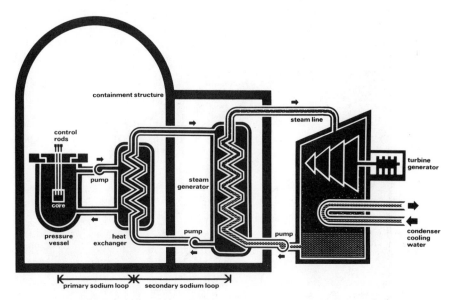

Fig. 10. A liquid-metal-cooled fast breeder reactor. (*Courtesy of the Atomic Industrial Forum.*)

assemblies, control assemblies, coolant circulation channels, and radiation shields are all contained in a reactor *pressure vessel* that is designed to withstand the enormous pressures of the coolant as well as to isolate the reactor core from the rest of the steam supply system. The pressure vessel has inlet and outlet nozzles for each primary coolant loop. The cap or head of the vessel can be removed for refueling and maintenance. The vessel is fabricated out of low-alloy steel to accommodate the high coolant pressures and temperatures and to withstand damage from the radiation generated in the core. The reactor pressure vessel, along with the rest of the nuclear steam supply system, is completely enclosed in a reactor containment structure that prevents the release of radioactivity to the environment in the event of a gross failure of the reactor coolant system. This nuclear island within the plant is usually a steel-lined concrete structure and contains not only the reactor itself, but the primary coolant system as well, including the primary pumps, steam generators, piping, and auxiliary systems. A boiling water reactor is shown in figure 12. Figure 13 shows the entire plant.

This brief technical description will suffice for our present discussion of nuclear power generation. Of equal importance in understanding the present debate over nuclear energy is the history of the development of nuclear power in this country.

Fig. 11. Reactor fuel assemblies. (*Courtesy of the General Electric Company.*)

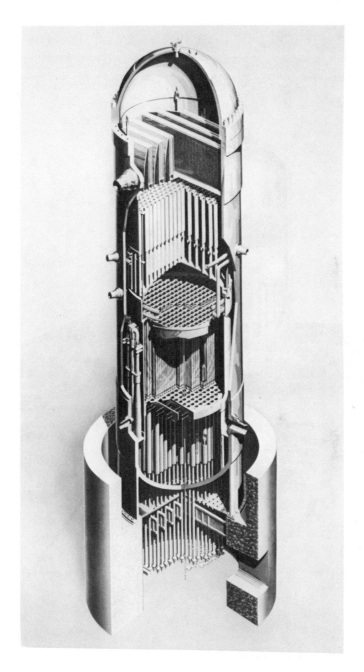

Fig. 12. A boiling water reactor. (*Courtesy of the General Electric Company.*)

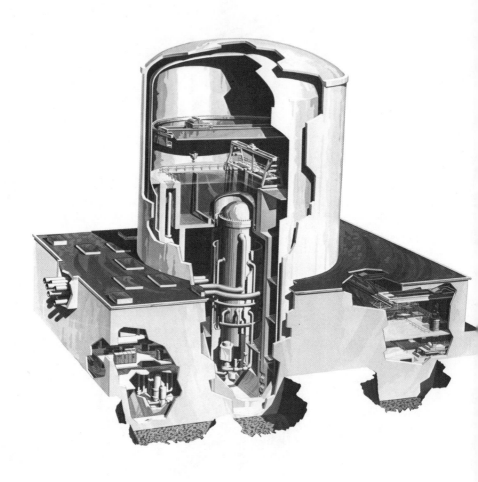

Fig. 13. A nuclear power plant based on a boiling water reactor. (*Courtesy of the General Electric Company.*)

The History of Nuclear Power Development

The fact that atomic energy was first developed and applied as an instrument of war cannot be overlooked in any discussion of nuclear power development. The awesome destructive power of nuclear weapons continues to tinge the emotional debate over nuclear power and to influence public attitudes toward this technology. The ambivalent potential of atomic energy for both war and peace has had an enormous impact on efforts to control this energy source. The first major legislation addressing nuclear technology, the Atomic Energy Act of 1946, voiced concern in its preface:

> The effect of the use of atomic energy for civilian purposes upon social, economic, and political structures of today cannot now be determined. It is reasonable to anticipate, however, that tapping this new source of energy will cause profound changes in our present way of life.[9]

To understand the development of nuclear power and the attitudes that now shape technical, political, and legal efforts to control it, it is useful to trace briefly its history. The dramatic story of the wartime development of atomic energy is a familiar part of twentieth-century history and has been told and retold in hundreds of sources. We pick up the story of nuclear power development in 1946, shortly after the end of World War II.

Since the early years of the atomic energy program, scientists had high hopes for peaceful applications of the atom. The responsibility for developing this potential was assigned to the newly formed Atomic Energy Commission (AEC) by the Atomic Energy Act of 1946. The primary goal was that of nuclear power, yet the difficulties that lay between this objective in 1946 and the first operation of a civilian power reactor some ten years later were real and complex. The whole field of knowledge on which reactor technology rested was strongly interlaced with nuclear weapons development. There were many questions of reactor safety and regulation to be answered. Since the 1946 act had given the commission absolute monopoly over nuclear materials, all of the early development work had to be conducted within the government program. Although the AEC recognized the importance of nuclear power development and wanted to initiate a major program in this area, it was not until 1949 that it convinced Congress to support the effort and organized its Reactor Research Division. Early progress toward Einstein's "almost certain" goal of atomic power was erratic, at best.

The basis for nuclear technology was laid during the war years. From Fermi's demonstration of the first critical chain reaction at the University of Chicago in 1942, the Manhattan Project progressed to large Hanford plutonium production reactors, and then to more sophisticated designs for research. By the end of the war a large variety of reactors had been built, operated, or studied. In 1946 the Manhattan Project launched the first atomic power program to develop an early commercial power reactor design, along with its naval and air force military reactor projects.[10] However, this program was reviewed by the AEC after its takeover in 1947 and eventually was halted because of growing pessimism concerning the possibility of achieving economically competitive nuclear power. A decision was made to concentrate all reactor work at the Argonne National Laboratory near Chicago, thus uprooting existing facilities at Oak Ridge, Tennessee. It was not until 1948 that a new program was formulated that involved activities with a materials testing reactor, a land-based submarine reactor, and an experimental breeder reactor, along with design work on a full-scale land-based power plant.

It was recognized at an early date that private industry must be allowed to enter the nuclear power field. But prior to 1950 little was done to attract the private sector. The 1946 Atomic Energy Act had created a government monopoly in atomic energy development. The government also controlled the reactor market since most of the early reactor designs were for military applications. The architects of the Atomic Energy Act had recognized the need to relax the government monopoly at a future date to allow for private development. Yet they failed to allow for the gradual entrance of private enterprise. The nation had yet to create a climate, both technical and legal, in which private nuclear power could develop.

Further complicating the transfer to private development was the ever-present private versus public power issue. The first AEC chairman David Lilienthal[11] felt that it was important that big business not be allowed to get a stranglehold on this great new natural resource "as they did in electrical power." Intense as this debate was, however, it only tended to obscure the real issues. Because of the immense capital investments and technical experience needed to enter the nuclear power field, government assistance and subsidy were mandatory if a private nuclear power industry was ever to be formed. To overcome private industry's fears of government monopoly and to introduce them to nuclear technology, the AEC issued an important statement in 1953 encouraging free competition and private investment in power reactor development while recognizing the government's responsibility for providing technical assistance in this venture. A further important step was taken with the 1954 Atomic Energy Act. This act allowed private ownership of nuclear materials and reactors and revised the patent laws to create higher personal incentives toward development; but the Eisenhower administration failed to follow up with necessary action, and in many respects simply slackened government efforts, thereby slowing progress toward a civilian nuclear power industry.

During the 1953–58 period the AEC made several more attempts to engage private industry in nuclear power. Known as the Five-Year Program, this effort consisted largely of small experimental reactors aimed at providing a basis for further technology development, although it included the authorization for a land-based civilian power reactor at Shippingport, Pennsylvania, which was financed in part by an electric utility, Duquesne Light Company, and was later to become the United States's first commercial power reactor. A Power Demonstration Reactor Program was launched in 1955, and several privately financed proposals were entertained and approved. These included the Yankee PWR plant in Rowe, Massachusetts, a small sodium-cooled reactor at Hallam, Nebraska, and the Enrico Fermi fast breeder reactor near Detroit, Michigan. At the same time three more plants were ordered by private industry: the Vallecitos BWR in California, the Dresden BWR near Chicago, and the Indian Point PWR in New York.

To stimulate industrial activity still further, the AEC initiated a second development program to encourage the construction of several other small-scale prototypes, including reactors cooled with organic coolants, heavy water, and helium. The general goal was to continue development of a given reactor type until its technology could be demonstrated or until it was established that it was an inferior design. Under this program the AEC offered to finance the reactor portion of any plant, while providing technical assistance and waiving fuel costs. As a final incentive, Congress passed in 1957 the Price-Anderson Act, which limited the liability of the operating utility in the event of a nuclear accident (an action we will discuss further in chapter 3). Thus, by the end of the 1950s, the American nuclear power program was broadened to include increased participation from private industry and appeared to have a sound technical foundation.

The evolution in nuclear reactor development from the primitive graphite-natural uranium pile constructed by Fermi to the enormous power reactors that generate much of the electricity used by the world today is an excellent example of the various stages involved in the progress of a new technology from scientific feasibility to economic viability. During the early period of nuclear power development at least nineteen different reactor types appeared to possess strong potential for commercial development and received detailed study.[12] Millions of dollars were invested in the development of each of these systems, and eleven of them reached the point where experimental reactors were built to demonstrate their technology. In many cases the shortcomings that eliminated a concept from further study did not appear until rather late in its development program. For example, the organic-liquid-cooled reactor looked extremely attractive until it was discovered in an experimental reactor prototype that the organic coolant decomposed rather rapidly under intense radiation in the reactor core. Most of the other reactor concepts similarly were abandoned as technical difficulties were encountered, although in many cases considerable amounts of money and manpower were invested. Development of most of these reactors was unsuccessful because of technical, economic, or political problems whose severity was underestimated or unknown during the early development stages.

The list of potential commercial reactor types was eventually narrowed to two light water reactor concepts: pressurized water reactors and boiling water reactors. Certainly a major factor in the success of the light water reactor program was the extensive technical experience acquired through the Naval Reactors Program that was based on water-cooled reactors. The first commercial power plant at Shippingport used a pressurized water reactor that was similar to that used in early nuclear submarines such as the Nautilus. Both Westinghouse and General Electric, in cooperation with the utility industry, built on an extensive experience gained through the Naval Reactors Program and the AEC Test Reactor Program (such as the Experimental Boiling Water Reactor) to build a series of demonstration plants, including the Yankee Rowe, Dresden I, and Indian Point I power plants. These plants demonstrated both the suitability of nuclear power for central station utility use and the capability of American industry to supply the necessary components. At the same time other reactor types were carried through to the demonstration plant stage, including the gas-cooled reactor (Peach Bottom), the sodium-cooled, graphite-moderated reactor (Hallam), and the liquid-metal-cooled fast breeder reactor (Enrico Fermi I).

When most of these demonstration plants went into operation in the early 1960s, the nuclear industry was already giving serious thought to the next generation of power reactors. These would be designed to be commercially viable and to compete economically with fossil-fuel-fired plants. Unfortunately the industry set unrealistic goals by demanding that such plants achieve power costs in the 6 to 7 mills per kilowatt-hour (mills/kwhr) range. Both General Electric and Westinghouse committed themselves to the construction of a number of introductory plants on a "turnkey" basis—that is, they signed agreements to provide the plants at a fixed price that would result in power costs at this level. But such plants never did produce the 6 mills/kwhr projected for them; the two reactor manufacturers eventually poured more than a billion dollars in unanticipated costs into the plants.[13] Nevertheless it was the commitment of the turnkey plants that transformed nuclear power from a series of costly single demonstration units to a commercially viable industry. It permitted the development of standardized engineering techniques and the buildup of the necessary engineering force, transforming the industry into more of an assembly line operation than a one-of-a-kind endeavor.

Gradual increases in the cost of fossil-fuel-generated electricity from 7 mills/kwhr in the early 1960s and 10 mills/kwhr by 1970 to its present level in excess of 20 mills/kwhr cast a much different light on the economic attractiveness of nuclear power. The economics of scale became more evident, and the capacity of nuclear plants was upgraded from several hundred to 1,000 to 1,300 megawatt units. Early milestones along the road to economic viability were the Oyster Creek and Nine Mile Point BWR power plants, which are usually regarded as the first commercially viable nuclear units, and the TVA Brown's Ferry plant, which contained three of the first 1,000 megawatt BWR units.

As the economic attractiveness of nuclear power became more evident during the late 1960s, utilities began to order plants in large numbers, and the industry built up the necessary manufacturing capacity to supply plants at a rate of forty to fifty per year. At this time roughly one of every two new power plants was nuclear.

Optimism for light water reactor technology became so strong that the AEC decided to concentrate its research effort on advanced reactor types such as the fast breeder reactor and leave further light water reactor development entirely to private industry. In retrospect this appears to have been an unfortunate decision, since just when light water reactors were being installed at a rapid rate, the agency responsible for their licensing and regulation decided to shift away from research on this technology, including research on the safety of light water reactors. Even so, the investment of public funds in reactor development was considerable and continues to be so. It is estimated that to date the federal government has invested some $3 billion in light water reactor development, including safety, fuel cycle, and supporting work, out of a total $9 billion for research and development on nuclear power.[14] (Most of this total has been used to support advanced concepts such as the breeder reactor and general research on materials, radiation effects, instrumentation, and other supporting technologies.)

As the nuclear power industry matured, it stimulated a variety of new patterns for the assessment and implementation of new technology. For example, the degree of safety regulation and evaluation required of nuclear plants set new standards for component fabrication and quality assurance in many industries. In fact the concern over proper separation between the regulation and the development of the new industry eventually led to the dissolution of the Atomic Energy Commission and its replacement by an independent Nuclear Regulatory Commission to monitor nuclear power safety.[15] The remaining development functions of the AEC were later assumed by the Department of Energy.

The nuclear power industry was forced to break ground in yet another area, that of environmental impact assessment. Prior to 1971 the primary responsibility for regulating nuclear plants, for issuing construction and operating licenses, rested with the Atomic Energy Commission, and the commission confined its regulation of such plants primarily to the areas of nuclear plant safety and radiation releases. In a landmark case involving the Calvert Cliffs nuclear plant in Maryland, the Supreme Court ruled that the AEC was also responsible for ensuring that the environmental impact of nuclear power plants was consistent with the National Environmental Policy Act (NEPA) of 1969.[16] Since that time the AEC and its successor, the Nuclear Regulatory Commission, have broadened their regulatory responsibilities to include not only questions of radiological safety, but environmental impact assessment as well. It also conducted legal studies to determine possible antitrust implications for the utility. Since there had been little experience in regulating any industry across such a broad spectrum, the AEC/NRC was forced to develop an entirely new set of regulatory standards, guidelines, and procedures.

As the industry matured, however, it encountered far more serious problems. The capital-intensive nature of nuclear power plants (some 80 percent of their generating costs can be attributed to construction and interest charges, only 15 percent to fuel charges) made the economic attractiveness of nuclear power extremely sensitive to the fluctuations of the economy, the availability of investment capital, inflation, shortages of materials, and so on. The cost of nuclear power plants rose dramatically during the early 1970s and continues to rise today at a rapid pace. The prospects for bringing nuclear power plants on-line for several hundred dollars per kilowatt capacity have faded to the status of wishful thinking as most utilities now project capital costs in excess of $1,000 per kilowatt for plants ordered during the next few years.

However, nuclear power has remained an economically attractive option in a relative sense, since the construction and fuel costs of fossil-fuel-fired plants have also escalated rapidly.[17] Today the construction costs of a coal-fired plant equipped with the necessary exhaust stack gas scrubbers are only slightly less than those of a comparable-size nuclear unit.[18] Nevertheless the dramatic escalation of capital costs, coupled with the economic recession that occurred in 1973–74, led to slowdowns and in some cases cancellations in the construction of plants already on order as electric utilities found it harder and harder to raise the enormous amounts of capital required to construct such plants.

Even so, the commitment to nuclear power generation by the late 1970s assumed staggering proportions.[19] By 1978 over 12 percent of the electricity generated in the United States was produced by nuclear power plants. Based on plants under construction at that time, this fraction will rise to 20 percent by 1985. Nuclear power plants in operation today represent a capital investment of about $20 billion; those under construction, an additional $75 billion. The foreign commitment to this new energy source paralleled that of the United States. Most European nations had been ordering and constructing nuclear units for several years because of their desire to free themselves from dependence on imported fossil fuels. In many nations this commitment to nuclear power generation exceeds that of the United States. (More detailed figures on the status of nuclear power as of 1978 are given in table 2 and figure 14.)

Perhaps the most serious problem faced by the nuclear power industry during recent years has been that of public acceptance. As the commitment to nuclear power has grown, so too has public opposition to this new technology. What was once regarded as a panacea for the ills of society is now viewed by many as "a last resort" at best,[20] and at worst, one of the most significant dangers in our society. One should not underestimate the importance of public

TABLE 2. Status of Nuclear Power Plant Commitments (1978)

	Number (and Gigawatts)	
	United States	World (non-U.S.)
Operating	71 (51)	151 (56)
Under construction	95 (103)	155 (127)
On order or planned	38 (44)	280 (252)
Total	204 (198)	586 (436)

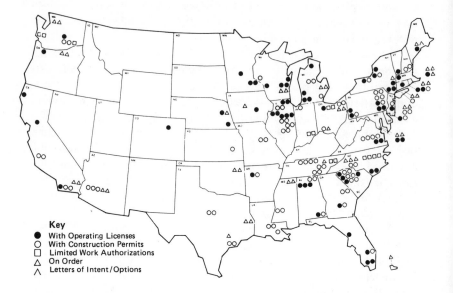

Fig. 14. Status of nuclear power plant construction in the United States as of January 1, 1978. (*Courtesy of the Atomic Industrial Forum.*)

acceptance of a new technology, for no matter how much effort or concern scientists and engineers invest in bringing a technology to fruition, and no matter how committed private industry and government may be to its implementation, in the long run the public will decide—and indeed, the public *should* decide—whether it will accept the technology as an integral part of its everyday life. For it alone will create the atmosphere that will either encourage or discourage the development and implementation of such technologies.

For that reason it is particularly important that we briefly examine the evolution of public opposition to nuclear power and try to identify the issues involved in the public debate concerning its deployment. It will provide an important lesson in how new technology is embraced or rejected by modern society.

The Nuclear Power Debate

The Gathering Storm

The roots of the present opposition to nuclear power can be traced back to the early days of the atomic energy program. A number of the scientists involved in the Manhattan Project were strongly opposed to the use of the first atomic weapons against Japan. This opposition to the development and deployment of nuclear weapons grew during the late 1940s and early 1950s. A particularly bitter debate[21] took place over the development of thermonuclear weapons (the hydrogen bomb). Although the decision was eventually made to proceed with the development of these weapons, organized opposition to nuclear weapons development continued to grow and eventually played a leading role in stimulating and approving a treaty banning atmospheric testing of nuclear weapons in the early 1960s. Scientific opposition to nuclear weapons development has continued to influence government policy on a number of significant issues, such as the decision to avoid massive deployment of antiballistic missiles and to push for more significant arms limitations in the Strategic Arms Limitation Treaty (SALT) talks.

It was natural that many of the same scientists who had played a leading role in opposing further nuclear weapons development during the decades following World War II would come to reexamine the peaceful role of atomic energy, to question whether the atom would really be the enormous benefit to society that had been promised. Although the vast majority of scientists, including many who actively opposed nuclear weapons, agreed that the potential benefits of nuclear power far outweighed its possible drawbacks, several were opposed to a massive nuclear power development program, even during the early 1960s.[22]

A number of events in the latter part of the sixties stimulated opposition to nuclear power. One important factor was the environmental movement and the growing awareness of the damage that man could cause to his environment. This was accompanied by heightened public involvement and reaction to technological change. It became quite popular to question the benefits of science and technology in general. For a time our society developed an almost antitechnology psychosis, viewing all technology as evil and basically harmful to man and his environment. The country was swept by a "return to the simple life" mood; grass roots movements tried to turn back the clock to a time when the air was clean and the cities were safe (although it was rarely recognized that the simple life of bygone years was also shorter and less comfortable, with little leisure time).[23]

This suspicion of technology was accompanied by a number of social upheavals (for example, Vietnam and Watergate) that led to a mood of questioning all social institutions. The public lost confidence in all established institutions and was generally willing to listen and give credence to any critics of the establishment, regardless of their qualifications. It was only natural that the growing nuclear power industry would become a prime target. After all, was not nuclear power simply one more scheme foisted on the public by big business and the federal government? Nuclear technology was particularly vulnerable since the rapid increase in electricity rates in the late 1960s and early 1970s created the impression that utilities were monoliths with myopic concerns for profit and little concern for public interest. Nuclear power became the symbol of big government and industry, of unfettered science and technology.

Hence the relatively token scientific opposition to nuclear power that had been present since the early days of its development was strengthened by new crusaders from the environmentalist movement as well as those in society who harbored suspicions of big business, or government, or the establishment, or technology (or almost anything else over which they did not have absolute personal control). And as the opposition to nuclear power began to swell, the character of this opposition began to change dramatically.[24] Many of those who had originally opposed nuclear power gradually reversed their attitudes as they learned more about the moderate risks and substantial benefits characterizing this new energy source. But they were rapidly replaced by a new breed of critics.

During the early years most critics of nuclear power had come directly from the scientific ranks (indeed, many of them had been directly involved in the wartime atomic energy program) and had tended to base their opposition on a direct knowledge of the scientific and technical facts involved in nuclear power generation. As the antinuclear power movement grew during the 1960s, it was joined by large numbers of critics with little scientific background or training. Certainly many concerned individuals in this movement recognized their severe limitations in evaluating issues that involved sophisticated technology such as nuclear power, and therefore they relied on the advice and interpretations of several scientists whose opposition to nuclear power was well known. Unfortunately the confusion and misunderstanding that resulted when well-intentioned laymen tried to interpret a scientific debate and reexpress it in a manner more acceptable to the public at large significantly changed the course of the debate away from technological issues and toward emotional issues that frequently were based on misconceptions and misinterpretations of scientific principles. Various new laws and regulations governing the implementation of a new technology (for example, the National Environmental Policy Act) further contributed to the confusion by setting up numerous public hearing procedures that were ready-made for legal maneuvering and delay. Hence the antinuclear power movement was taken over largely by lawyers, and the original scientific dialogue was replaced by legal maneuvering and political oratory, which usually had little relevance to the implementation of nuclear technology.

The shift of the antinuclear power movement from a scientific to a political orientation over the past decade is apparent in the nature of the issues raised by critics of nuclear power. The earliest arguments were those concerned with the low-level releases of radioactivity from nuclear plants. Several scientists[25] received great public attention by proclaiming that the low-level radiation released during normal operation of nuclear power plants could cause thousands of cases of cancer every year. These claims were quickly and thoroughly discredited by the radiological health physics community.[26] Nevertheless the public concern generated by these frightening (if incorrect) statements forced the federal government to lower the allowable limits on public radiation exposures from nuclear plant radioactivity emissions to a level a hundred times below that applied to any other sources, such as medical or industrial—even though there was no evidence for the necessity of these lower limits.[27] (We will consider radioactivity emissions from nuclear power plants in detail in chapter 3.)

It has taken the scientific community a number of years to repair the damage to public confidence caused by such irresponsible claims. Gradually the public has become aware that the routine emissions from nuclear plants give rise to public radiation exposures a thousand times smaller than those from other radiation sources in our environment, such as medical X rays, natural radioactivity, and fallout from nuclear weapons testing.

As the debate over low-level radiation waned, opponents of nuclear power turned their attention to nuclear reactor safety. Although commercial nuclear power plants have never experienced an accident in which a member of the public was harmed, there is always a remote possibility that such an accident might occur. Nuclear critics seized on several preliminary studies by the AEC[28] to estimate the worst possible consequences of a catastrophic reactor accident during the 1950s in their effort to persuade the public of the enormous dangers of nuclear power. Of course, the AEC was partly to blame in this instance since it had deferred research on light water reactor safety during the 1960s just when light water reactors were being deployed commercially on a massive scale by electric utilities throughout the country.

A number of scientists expressed genuine concern about the lack of experimental data on the performance of nuclear plant safety systems.[29] This criticism had a positive effect since it pressured the nuclear power industry and the federal government into reemphasizing and accelerating their research program on reactor safety. However, when coupled with incidents such as the fire at the Browns Ferry plant in 1975 and the accident at the Three Mile Island plant in 1979, this criticism also intensified public fears of nuclear power technology. The public debate over nuclear safety shifted into high gear when various consumer advocate groups seized on nuclear reactor safety as their next target.[30] These groups began a series of highly emotional, yet carefully organized, attempts to pass legislation or popular referenda to prohibit the construction of nuclear power plants on the basis of their presumed dangers. As these groups entered what had been primarily a technical debate, scientific reason seems to have departed. Today we are faced with a highly emotional battle between groups with limited technical backgrounds, little understanding of the issues involved in nuclear reactor safety, but exceptional skills in gaining the attention of both the public and their elected representatives in government.

Many opponents of nuclear power have escalated their disaffection from specific concerns about safety or the environment to the more general conclusion that nuclear technology itself is somehow immoral and therefore should be eliminated. This has led to some amazing public actions. The National Council of Churches condemned atomic element 94 (plutonium) on supposedly moral grounds while opposing nuclear power as a major hazard to our society.[31] It has become fashionable to refer to the implementation of nuclear power as a Faustian bargain, suggesting that we have somehow made a pact with the devil by developing a technology with such awesome hazards.[32] The mixture of moral questions into what had been a technical debate has only confused the issues even further.

Most recently nuclear critics have broadened their aim to encompass the entire nuclear industry rather than simply the power plants themselves. A variety of concerns have been expressed about the adequacy of domestic uranium ore supplies, the reliability and economics of nuclear plant operation, and our capability of producing net energy from nuclear power with the enormous investment required by plant construction and fuel processing. But most criticism has been directed at the tail end of the nuclear fuel cycle,[33] the reprocessing of fuel discharged from nuclear power plants, the possible use of the plutonium separated out from spent fuel in nuclear weapons (either by terrorists or other nations), and the disposal of the radioactive waste produced by nuclear plants. All of these issues will be examined in some detail in later chapters.

As the composition of the movement opposing nuclear power changed, so too did the nature of the criticism of this new technology. The token opposition voiced by a handful of scientists that had been directly involved in the early atomic energy program gradually gave way to a large and active antinuclear movement composed of nonnuclear scientists with backgrounds in the natural, medical, or social sciences, along with a number of highly concerned, but relatively uninformed laymen. As the opposition to nuclear power became more highly publicized, it became rather fashionable for those opposed to society as we know it to include opposition to nuclear power as part of their own particular crusades. This was particularly true of the environmental movement. Many groups have embraced claims that nuclear power will do severe damage to the environment and have mounted massive campaigns to block nuclear plant construction.[34] The opportunities for these activities are numerous because of the openness of the nuclear licensing and regulation process and the wide provisions for court review of administrative decisions. Essentially every nuclear plant under construction quickly found itself enmeshed in a tangle of litigation on first one issue and then another.

The battle over nuclear power also quickly attracted large numbers of political activists whose primary target was not so much nuclear power as the entire fabric of our society. Because of the stringent government control over nuclear energy development and the awkwardness with which federal agencies such as the AEC interacted with the public, nuclear power presented a rather vulnerable target. For many of those in the antinuclear movement, opposition to nuclear power was merely a vehicle for forcing major social change; it was only the first of many battles to be fought on this front. In many cases, concerns for nuclear reactor safety, environmental impact, or economic viability were secondary. Rather the opposition to nuclear power was intended to deny our society a means for satisfying its growing appetite for energy, thereby forcing massive conservation efforts that would lead to a less energy-intensive way of life. Certainly those in opposition to nuclear power were far from united in either the degree or the nature of their concerns.

Today, opposition to nuclear power has become fashionable as a mechanism of political protest. And as the various legal proceedings against nuclear plant construction are exhausted (frequently in the Supreme Court itself), protesters are turning more and more frequently to the techniques of civil disobedience learned so well during the Vietnam War.[35] Each nuclear plant construction project is now accompanied by an organized group of protesters committed to blocking the construction or operation of the plant at almost any cost (for example, the Clamshell Alliance for the Seabrook Plant in New Hampshire, the Abalone Alliance for the Diablo Canyon plant in California, and the Arbor Alliance for the plants in Michigan).

Issues in the Nuclear Debate
As the nuclear power industry has matured from scientific feasibility to economic viability (and perhaps to a future status of technological *inevitability* should the development of other long-range energy alternatives prove unsuccessful), so too has the public controversy grown over the role that nuclear power should play in our society. Why did this concern arise, and why does it receive so much public attention?

Certainly one reason is that nuclear power has become a very prominent target. Since 1966, nuclear power plants have presented a viable and economic source of electric power. The energy crisis and the shortage of fluid fossil fuels has only increased the incentive to switch to nuclear plants for generating electricity. For example, during 1978 the electric generating costs for nuclear power were some 20 percent less than those for coal and over 160 percent less than those for oil, a savings of some $2 billion (with some seventy nuclear plants in operation).[36] Electric utilities are convinced that nuclear power presents a safe and clean source of electric energy that is capable of massive implementation today. Since they feel that nuclear power is cheaper, cleaner, safer, and of greater immediate potential than any other presently viable alternative, they have chosen to make staggering commitments to nuclear plant construction (having already invested $20 billion with an additional commitment of $75 billion).[37]

A second factor that contributes to the ease with which opposition can be voiced against nuclear power involves the ready availability of information relating to nuclear plant safety and environmental impact. These data have been accumulated over the past several decades by the federal government and by private industry, and they stand out in contrast to the paucity of data about alternative methods of electric power generation. This material provides ample fuel for those who would selectively attack this new technology.

The nuclear licensing and regulation process also contributes to this trend since it provides numerous opportunities for public hearings and intervention. Coupled with the variety of avenues for litigation and court review, licensing and regulation practices tend to keep opposition to nuclear power constantly before the public eye.

Finally, and probably most significant, nuclear power is still rather mysterious to the public. The haunting memory of Hiroshima continues to hang over our society. We can see evidence of public fears even in the language used to describe nuclear power: "invisible" radiation, radioactive "waste" disposal, nuclear sabotage and terrorism, plutonium. This mystery, these fears, contribute to the emotional and frequently irrational debate over this energy source.

So what are the issues in the nuclear power debate? In table 3 we have listed and compared some of the various pros and cons of nuclear power. In the pro column we have listed the claims that nuclear power is cheaper, safer, and cleaner than other alternatives and that it is available for immediate and massive implementation. The proponents of nuclear power claim that the past decade of nuclear plant operation has demonstrated these advantages.[38] Nuclear power plants do generate electric power more cheaply than fossil-fueled plants. During their operation there is essentially no release of combustion products to the environment. Furthermore in almost twenty-five years of commercial reactor experience, there has yet to be an accident (including Three Mile Island) that has had any measurable effect on public health. Finally, proponents of

TABLE 3. The Pros and Cons of Nuclear Power

Pros	Cons *(popular press)*
Cheaper	Nuclear plant safety
Safer	Low-level radiation releases
Cleaner	Waste heat discharges
Available now	Radioactive waste disposal
Necessary to meet demand	Sabotage and nuclear theft
Sizable fuel reserves	Nuclear weapons proliferation
	Economics, reliability, energy payback
	Cons (subconscious)
	Legacy of Hiroshima
	Nuclear = strange, new, invisible
	Natural suspicion of technology
	Antiestablishment
	Means to force conversation
	Other problems
	Public acceptance
	Inconsistency (or lack) of federal policy
	Complexities of federal regulations
	Financing energy development
	International aspects

nuclear power assert that this technology is necessary if we are to meet the electrical demands of our society during the next several decades, and that sufficient domestic uranium reserves are available to make the present type of light water reactor a viable source of energy until well after the turn of the century. With the introduction of the fast breeder reactor, nuclear power generation will represent essentially an unlimited source of energy.

We have separated the arguments against nuclear power into three groups. Most of the items in the first group have received intense public exposure in the media. These include concern over nuclear reactor safety and low-level radiation releases, the environmental impact of both nuclear plants and their associated fuel cycle, the disposal of the radioactive waste produced by such plants, the possibilities of sabotage of nuclear power plants and theft of nuclear materials that might be suitable for nuclear weapons fabrication, the degree to which nuclear power accentuates the international proliferation of nuclear weapons, and concerns about the economics, reliability, and the energy efficiency of nuclear power. This is certainly an imposing list.

A number of subconscious factors also are involved in the debate. Nuclear power continues to be burdened by the legacy of Hiroshima. Many in our society are driven by a suppressed guilt complex engendered by the role that our nation played in the development and military use of nuclear weapons. Their opposition to nuclear power may be a manifestation of their deep-rooted horror and revulsion with nuclear weapons. They act as if they believe that dismantling the nuclear power industry will return us to a world without the bomb, without the possibility of nuclear war. Certainly the emphasis of the early atomic energy program on military applications has contributed to the public's view of all nuclear technology as a mysterious and sinister force whose destructive potential far outweighs any peaceful benefits. Although this emphasis has changed during the past decade, the aura of the weapons program remains, and nuclear power is far more likely to trigger the image of a mushroom cloud than that of a clean, efficient power plant generating much needed electric energy.

There are other subconscious elements involved in the opposition to nuclear power. One segment of the antinuclear movement views any technology with great suspicion and wants to force our society back to a simpler way of life in which dependence on technology is minimized. To these individuals, opposition to nuclear power is merely the beachhead in a more general battle against all technology.[39] Many scientists share this natural suspicion of technology with the layman. In this sense one must distinguish between scientists who are usually concerned with idealized studies of fundamental scientific principles and engineers who must deal with highly complex applications of science and technology to society. Since few scientists have experience or training in engineering design or applications, they frequently find it difficult to believe that anyone can deal with the myriad technical problems and conflicting goals of real and complex systems involved in practical applications that are routinely dealt with by engineers.

We must not dismiss opposition to nuclear power as simply an emotional manifestation of suppressed fear or guilt concerning nuclear weapons or a general reaction against technology, however. Some very real problems must be overcome if nuclear power is to realize its potential. Public acceptance, or the lack thereof, is a major barrier to massive deployment of nuclear power, both in this country and abroad. Moreover the process by which nuclear power is regulated, licensed, and controlled continues to flounder in a mass of red tape and bureaucratic delay. The rapid escalation of the already staggering cost of central station power plants—both nuclear and fossil-fueled—may well exceed the ability of our society to finance such construction from the private sector. Certainly, too, the international aspects associated with the spread of nuclear technology, particularly those associated with the nuclear fuel cycle, are related to the proliferation of nuclear weapons capability and require immediate and serious attention.

We will consider all of these points in later chapters and provide readers with enough technical background and information to perform their own evaluations of nuclear power in our society. But, first, we offer several cautions to be kept in mind when evaluating the nuclear power debate.

First, nuclear power generation is an extremely broad subject. Any individual can claim expertise in only a narrow subfield. Hence, one should examine carefully the credentials and the qualifications of those who make statements in the nuclear power debate. For example, one should be cautious about statements concerning nuclear reactor safety made by lawyers or utility executives, just as one would discount statements on economics or legal matters made by engineers. Indeed, one should take assertions about engineering technology from scientists with a grain of salt, since scientists are usually unaccustomed to dealing with complex systems, the dirty problems involving practical applications that are routinely faced by engineers. One should also be wary of broad generalizations about any complex new technology such as nuclear power.

Credibility is always a problem in an emotional debate. This is particularly true in the debate over nuclear power, since most individuals with extensive experience in nuclear power generation have strong ties to either the nuclear power industry or the federal government. It is natural to be suspicious of anything big and to transfer this suspicion to those associated with the atomic energy program. But we must remember that the electric utilities, the federal government, and those proponents of nuclear power from the universities are not necessarily stupid or out to mislead the public intentionally. Rather, they appear to be making a conscientious and honest effort to arrive at correct decisions about our future needs for energy. These individuals will be affected by these decisions in a personal manner as members of the public. They, too, have families. They, too, are concerned about questions of safety and the environment. And, interestingly enough, few proponents are likely to reap vast fortunes from the implementation of nuclear power, in sharp contrast to many attorneys who stand to benefit from enormous legal fees in the various lawsuits and public intervention cases conducted against the nuclear power industry. Proponents of nuclear power appear to be genuinely concerned about man's future.

Finally, the technical debate over the decision to implement nuclear power should be clearly separated from moral or emotional issues. An important principle to keep in mind when evaluating such debates is that of scientific integrity. When a scientist discusses an issue, he is obligated to present all known aspects of a subject, not simply those aspects that support his own beliefs. That is to say, if he is aware of technical facts that would tend to counter his argument, he is obligated to present those facts in his discussion. This contrasts with the advocate system of law in which an individual presents only those facts that support his side of a case. This is particularly important since so many lawyers have entered the debate over nuclear power, changing its nature in a significant way. Unfortunately, when the advocate system is applied to science, it tends to throw out scientific integrity and introduce pseudoscience.[40] This has confused the public.

The importance of responsible criticism during the development of any new technology cannot be overstressed. During the development of nuclear power, numerous responsible critics stimulated changes in the direction of the program and additional research, and pointed out problems overlooked by others. There is a real danger, however, that such responsible criticism will be overwhelmed by the emotional outbursts of those with only a marginal understanding of the technical issues, but with almost evangelistic zeal for alerting society to the enormous dangers and follies of nuclear power.

The role of the layman in influencing technical decisions is extremely important, since public funds, issues involving the environment, political and societal effects, and long term public welfare are frequently at stake. Some means must be found to enable the layman to raise legitimate questions without paralyzing efforts to obtain valid and supportable answers. Unfortunately, the complexity of the issues is frequently beyond those who have not had training, education, or experience in the field, and there is little wonder that the layman frequently gets confused.

Certainly some exposure to the technical facts is necessary if the public is to make correct decisions. Fortunately in many cases a few basic concepts, combined with a knowledge of sources of further information, can go a long way. It is possible to learn how to ask the right questions of those debating a particular technical issue, to distinguish the expert from the charlatan, and to identify individuals with valid concerns. A primary goal of this book is to provide readers with sufficient technical background and sources of information so that they can rationally examine the nuclear power issue themselves.

3

Nuclear Power: A Viable Technology?

For a technology to achieve social viability, the capability of massive implementation, it must meet the fourfold tests of resource availability, acceptable public risks, minimal environmental impact, and economic viability. These are also the criteria that should be used in comparing the suitability of one form of energy production with another.

It is evident that nuclear power is presently a viable technology since it generates some 12 percent of the electricity used in this country and a comparable amount in most of the other industrialized nations of the world.[1] There can be no argument on this score. However, we must address the more difficult question of whether nuclear power can *remain* viable as an energy source. The present debate over nuclear power has focused on the future role of this technology in meeting man's energy needs.

Resource Availability

All present nuclear power reactors use uranium as a primary fuel. Most of the uranium ore mined in the United States comes from the sedimentary sandstone and mudstone deposits of the Colorado Plateau, the Wyoming Basin, and the Gulf Coastal Plain. This ore yields about 0.1 to 1 percent U_3O_8, in contrast to the pitchblende deposits found in Canada, Czechoslovakia, and central Africa that can yield up to 20 percent U_3O_8. Low-grade uranium concentrations occur in the Chattanooga shales of eastern Tennessee (up to 60 parts per million) and the Conway granite formations (from 20 to 40 parts per million). Very low uranium concentrations (roughly 0.003 parts per million) also occur in seawater.[2]

Of most direct concern are the relatively high-grade uranium resources that can be mined at a cost of less than $50 per pound of U_3O_8. Present government estimates place known reserves at 890,000 tons, probable reserves at 1,395,000 tons, and possible or speculative resources at an additional 2,080,000 tons.[3] The significance of these estimates becomes apparent when it is recognized that a 1,000 megawatt nuclear plant will utilize some 5,000 tons of uranium during its thirty-year operating lifetime. Therefore the present commitment of 208 power plants will require some 1,040,000 tons of uranium ore, a quantity somewhat larger than the present known reserves. The 400 to 500 gigawatts of nuclear power forecast by the Department of Energy for the year 2000 would require some 2,000,000 tons, roughly the total of our known and probable domestic reserves. Therefore there is considerable uncertainty about the adequacy of domestic uranium resources to support conventional, light water reactor nuclear power plants constructed after the turn of the century.

The magnitude of our domestic uranium resources depends on the amount we are willing to invest in mining low-concentration deposits such as the Chattanooga shales or Conway granite formations. The low concentrations of these deposits not only imply large ore costs (in excess of $125 a pound), but also extensive mining and milling operations. In fact, the energy concentration of these ores approaches that of coal. There is some doubt that the energy investment required for extracting uranium from low-concentration deposits or seawater could be repaid by utilization of these resources in light water reactors.

The significance of these estimates of uranium resources depends sensitively on the reactor type. For example, thorium can also be used as a fuel (actually, a fertile material) in advanced converter reactors. However, thorium-fueled reactors also require small amounts of U-235 or plutonium to boost the neutron multiplication of the fuel to sustain a critical chain reaction. Experience in the design and operation of such reactors is limited, and the magnitude of thorium resources is not adequately known, although probable reserves are estimated at only about 500,000 tons.[4]

The adequacy of uranium resources changes abruptly with the introduction of the fast breeder reactor. By converting uranium-238 into the fissile element plutonium, this reactor type can extract over fifty times more energy out of uranium ore than light water reactors. There is sufficient uranium in the form of U-238 in the tails stockpiles at uranium enrichment plants to supply a fast breeder reactor economy for hundreds of years. In fact there would be no need to mine additional uranium for this period. Furthermore the relative insensitivity of the breeder to uranium ore costs would allow the exploitation of low-concentration deposits such as shales, granites, and even possibly seawater.

It should be apparent that the estimate of domestic uranium resources will influence rather heavily this nation's decision to proceed with the development of the fast breeder reactor. It is interesting to note that while essentially all domestic uranium data are supplied by a single source,[5] the interpretation of these data has led to marked differences in the estimates of these resources.[6] To support its recommendation that the development of breeder reactors be slowed, the Carter administration relied on optimistic estimates[7] of uranium resources that would supply light water reactors until well into the next century. However, these estimates are in sharp conflict with those of other scientific groups[8] that tend to place reserves as sufficient to supply only about 400 gigawatts of installed capacity.

Nuclear Power Plant Safety

What Can Happen?
How safe is nuclear power? What risks does this technology pose to our society? Of most concern has been the safety of the nuclear power plants themselves, particularly in view of incidents such as the Three Mile Island accident in 1979. To study nuclear plant safety, we will first consider situations in which a nuclear power plant is subjected to abnormal operating conditions caused, for example, by component malfunction, operator error, sabotage, or a host of other events that could lead to a nuclear reactor accident. The principal safety concerns involved in nuclear reactor operation do not arise because of the possibility of a nuclear explosion. Nuclear reactors cannot explode like atomic bombs because they are fundamentally different. Bombs require that highly concentrated U-235 or plutonium be rapidly assembled into a supercritical configuration. In light water reactors the fissile concentration (enrichment) of about 3 percent is far too dilute to allow for an explosive chain reaction. Furthermore negative feedback mechanisms in reactors turn off the chain reaction automatically if the power level increases substantially.

The principal concern in nuclear reactor safety is the large inventory of radioactive material that accumulates in the reactor fuel during operation.[9] An operating power reactor builds up an enormous quantity of radioactive fission products in its fuel, along with other radioactive materials produced by neutron bombardment. As long as this radioactivity remains in the fuel, it represents no immediate danger. However, should it be released and dispersed in the atmosphere in populated areas, it would be a significant hazard. Nuclear reactors must be designed so that under no credible—or even incredible—operating situation could significant quantities of radioactivity be released from the reactor core. To achieve this assurance, not only must the nuclear reactor and coolant system be carefully designed against every imaginable accident, but also auxiliary systems must be incorporated into the design to ensure core integrity in the event that such accidents should occur.

The containment of radioactive fission products is accomplished by designing into a nuclear power plant a series of physical barriers that inhibit or prevent the release of fission products or other radioactive material (see fig. 15). The first line of defense is the ceramic fuel pellet itself, which entrains most of the nongaseous fission products and greatly inhibits the diffusion of gaseous fission products out of the fuel. The fuel pellets are contained in metallic tubes or cladding of zirconium or stainless steel that retain even the gaseous fission products that build up in the gap between the fuel pellet surface and the cladding tube. The fuel elements are contained within a steel pressure vessel 20 centimeters thick that serves as a third barrier to fission product release. The primary coolant loop piping is 8 to 10 centimeters thick, and the coolant water itself is continuously circulated through filtering traps to separate out any radioactive material. The pressure vessel is surrounded by concrete

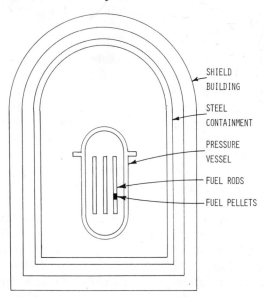

SHIELD
BUILDING

STEEL
CONTAINMENT

PRESSURE
VESSEL

FUEL RODS

FUEL PELLETS

Fig. 15. Physical barriers to fission product release.

shielding 2 to 3 meters thick in a containment building of concrete walls 1 meter thick lined with a 10 centimeter leak-tight steel shell that prevents the release of radioactivity in the event of a major rupture of the primary coolant system. The plant itself is contained in an exclusion area over which the operating utility has access control and which separates the plant from the public. Finally the plant site is located intentionally in a low-population zone some distance from any major population center.

Still other precautions are taken to ensure nuclear reactor safety. Major lines of defense include (1) *quality assurance* to guarantee that all components of the plant have been manufactured and assembled to required design specifications; (2) highly redundant and diverse *safety systems* designed to protect against abnormal operating conditions; (3) *engineered safety systems* designed to protect against the consequences of highly unlikely but potentially catastrophic accidents, such as loss of coolant, equipment failures, human error, and severe natural events—earthquakes, tornadoes, floods. This approach to nuclear plant safety is sometimes referred to as *defense in depth*. It implies that nuclear engineers must do everything possible to prevent accidents from happening through conservative design and safety systems. Then to cover the possibility that some systems will not work as intended, engineers must add on so-called engineered safety systems to minimize the consequences of any accident that might occur. All of these features are then augmented by complete and detailed testing and inspection procedures.

Consequences of hypothetical accidents are carefully analyzed and factored into plant design to protect the public in the event of such accidents. This design process is applied to ever more improbable events until a point is reached at which it is agreed by both designers and regulators that the situation assumed is impossible or incredible. The extremely unlikely accident that is just short of impossible is termed the *design basis accident*. A nuclear power plant must then be designed with sufficient safety margin to withstand the design basis accident without endangering the public. This then provides assurance that the plant has a design margin to withstand any lesser accident that might occur.

What types of accidents do reactor engineers consider in the design of safety systems? We have noted that most radioactive fission products produced during reactor operation are entrained in the ceramic fuel pellets. The only way this radioactivity can be released in massive quantities is for the pellets to melt. Hence the most serious accident postulated for a nuclear power plant is loss of cooling, which might lead to a reactor core *meltdown*.

The typical design basis accident for light water reactors involves a massive rupture of the primary coolant system. Since the water coolant is at extremely high temperature (300°C) and pressure (150 atmospheres), any rupture in the high-pressure coolant system will cause the water to flash into steam and blow out of the leak. The loss of moderation immediately shuts the chain reaction down, but the residual decay heat from radioactive fission products would tend to raise fuel temperature quite rapidly to its melting point (within thirty seconds), leading to clad failure and release of fission products from the primary coolant system unless auxiliary cooling is provided.

In the absence of cooling, the reactor core could melt in roughly thirty minutes and slump in a molten mass to the bottom of the reactor vessel. In several hours it could then melt through the vessel and containment building floor and release significant amounts of radioactive fission products to the environment. (Occasionally one hears reference to the China syndrome in which the molten fuel mass is imagined to continue to melt its way down into the earth. Studies have indicated that if the molten core were to melt through the concrete floor of the containment building, it would come to rest in the soil or bedrock several meters below the building foundation.)

Let us describe the *loss of coolant accident*[10] (LOCA) as postulated by the devious minds that dream up such scenarios to challenge the sanity of reactor engineers. One assumes that the reactor has been operating at full design power for some time when a double-ended fracture of the cold leg of the primary coolant piping occurs. That is, one assumes that the pipe bringing cold water into the reactor core breaks and is displaced in such a way that all flow in the pipe is discharged. The coolant in the pressure vessel rapidly depressurizes and blows out from both ends of the break as a mixture of water and steam. The voiding of the coolant from the core is referred to as the *blowdown* phase of the loss of coolant accident. The reactor goes subcritical as soon as significant boiling occurs since the reduction in the coolant density corresponds to a decrease in moderation.

The decay heat generated in the core continues to be substantial (about 5 percent of the operating power level) and will lead to a rapid rise in fuel element cladding temperature unless auxiliary cooling is provided. To this end the nuclear steam supply system is equipped with an *emergency core cooling system* (ECCS) designed to protect against fuel element melting and failure. Both active and passive systems are used. Large tanks of borated water called accumulators are maintained at pressures somewhat below the operating system pressure so that, in the event of a loss of coolant accident, the water from the accumulators will be discharged into the reactor vessel when the system pressure drops below that of the accumulator tanks. These devices operate in an entirely passive manner so that no separate control device is required to activate them. The emergency core cooling system also contains active systems utilizing both low- and high-pressure coolant injection pumps. The high-pressure injection pumps provide the cooling during an accident resulting from small area breaks. The low-pressure injection pumps are much higher capacity and are intended for long-term cooling during a large loss of coolant accident. It is the task of nuclear design to demonstrate that, following the accident, the fuel clad temperature is maintained below a critical limit (for zirconium this limit is taken as 1200°C) by the emergency system.[11]

Nuclear power plants are equipped with numerous other engineered safeguards. The containment systems themselves are engineered safety systems since they are designed to contain the contents of the primary coolant loop in the event of a rupture. In a pressurized water reactor the primary containment is provided by the containment building itself, which is usually designed to withstand an overpressure of some 4 atmospheres resulting from a failure of the primary coolant loop. Such containment structures are equipped with ice-condenser and spray systems to decrease these pressures. In a boiling water reactor plant the primary containment is provided by a steel-lined concrete bottle known as a drywell that contains the reactor pressure vessel. Below this drywell is a pressure suppression pool that acts as a steam dump in the event that the main steam valves to the turbine building would have to be closed. There is also a secondary containment provided by the reactor building itself, which is airtight, although not designed to contain high pressures.

Engineered safety systems designed to provide emergency core cooling or the containment of radioactive materials in the event of a serious reactor accident play a critical role in nuclear reactor safety. Since these systems are designed to protect the public in the event of a catastrophic accident, it would be extremely difficult and awkward to subject a commercial-sized power reactor to accident conditions such as a loss of coolant event just to test a particular safety system. Therefore the Department of Energy has constructed several experimental facilities to test safety components such as the emergency core cooling system.

The largest such facility is the Loss of Flow Test or LOFT program in Idaho. This consists of a scaled-down nuclear reactor and coolant system that can simulate the behavior of a pressurized water reactor steam supply system under accident conditions. The LOFT experiment has been constructed so that it can examine a large variety of possible loss of coolant accidents and test various components of the emergency core cooling system. These experiments also serve to test the predictions of analytical models used in the design and analysis of nuclear power plants. After a series of successful tests in a cold configuration with the reactor subcritical, full-power, hot testing of emergency core cooling systems commenced in late 1978. Preliminary test data indicate that these systems are extremely reliable and effective in providing core cooling in the event of a loss of coolant accident, and that, if anything, the emergency core cooling system design tends to be somewhat conservative.[12] However the thorough analysis of this engineering data is complex, and final results will probably not be available for several years.

Yet despite the care taken in nuclear power plant design, in the implementation of the defense-in-depth principle, and in the provision and extensive testing of engineered safety systems, there will inevitably be accidents in nuclear power plants. Therefore nuclear plant design must anticipate such incidents and provide adequate mechanisms to protect the public.

Although the safety record of nuclear power generation in the United States has been exceptional, there have nevertheless been several serious accidents.[13] However in all cases to date, the engineered safety systems have functioned as designed, and no harm to members of the public has resulted from these incidents. For example, in 1966 an accident occurred in the Enrico Fermi experimental breeder reactor near Detroit, Michigan, that resulted in blockage of coolant flow and partial melting of several fuel elements. Although significant reactor damage occurred, no radioactivity was released from containment, and danger to the public was minimal.

A more serious accident occurred in 1975 at the large Browns Ferry nuclear power plant in northern Alabama. In this instance a plant inspector started a fire in the plant that burned through a number of control cables before it could be extinguished. However, no direct damage to the nuclear reactor in the plant occurred, and there was no radiological danger.

Certainly the most serious plant accident to date was that which occurred at the Three Mile Island station near Harrisburg, Pennsylvania, in spring of 1979.[14] A combination of equipment failure, design deficiencies, and human error led to a loss of coolant accident. Although there were no casualties, the accident did result in extensive core damage and the release of some radioactivity off-site. But perhaps of more significance was the damage this accident caused to public support of future nuclear power development in this country as it intensified fears about the safety of nuclear power plants.

The accident occurred in the second pressurized water reactor unit of the plant. At 4 A.M. on March 28, while operating at 98 percent rated power, the pump in the secondary or steam loop which pumps water condensate from the main steam condenser stopped operating. When the feedwater pumps that circulate this water back to the steam generators lost suction, they automatically shut down. The loss of feedwater caused an automatic shutdown of the turbine. Three auxiliary feedwater pumps automatically went into operation, but were unable to pump water into the steam generators because connecting valves had been inadvertently left closed during an earlier maintenance operation.

Deprived of feedwater flow, the secondary side of the steam generators began to boil dry. But more significantly, the source of cooling was now lost for the primary coolant loops that passed through the reactor, the pressure in the primary loops began to increase as the primary coolant water continued to absorb heat from the reactor core. This caused a pressure relief valve on the pressurizer to open, discharging primary cooling water into a discharge tank in the containment building. Shortly thereafter, the increasing pressure triggered an automatic shutdown of the reactor. As the fission chain reaction subsided, the major source of heat in the primary loop disappeared, and the coolant pressure began to drop.

At this point several additional failures occurred that turned what had been a relatively minor operational malfunction into a serious accident. The pressure relief valve in the primary loop stuck in the open position, leading to a slow loss of coolant and also overflowing the pressure relief discharge tank, spilling primary cooling water onto the floor of the containment building. A sump pump that was automatically activated began to pump the spilled primary coolant water over into an adjacent building (the auxiliary building) of the plant.

As the primary system pressure dropped below operating levels due to the loss of cooling water through the open pressure relief valve, the emergency core cooling system automatically went into operation to provide reactor core cooling, just as it was designed to function. At this point human error entered the picture when the plant operators failed to notice the loss of coolant through the stuck relief valve and made a decision to turn off the emergency core cooling system in the mistaken belief that the primary coolant loop had filled solid with liquid water. This action led to a further decrease in primary coolant pressure which resulted in the formation of steam bubbles in the primary loop. This mixture of water and steam began to degrade the operation of the primary coolant pumps (pump cavitation), and the resulting pump vibration convinced the operators that the primary pumps should be turned off and that an effort should be made to establish cooling of the core by natural convection. However, the steam voids in the primary loop prevented natural circulation cooling, and the subsequent overheating of the core due to fission product radioactive decay heat produced serious core damage before the situation was fully understood.

A significant fraction of the core experienced fuel element failure, releasing radioactive fission products into the coolant and blocking many of the coolant flow channels in the central part of the core. The cooling water containing radioactive fission products then spilled through the open relief valve and discharge tank onto the floor of the containment building, and the sump pump transferred some of this radioactive water into the adjacent auxiliary building. Since this latter building was not designed to be leak tight, some of this radioactivity escaped through the building ventilation system into the atmosphere. This was the principal source of radioactivity release during the accident.

Although no fuel melting occurred in the reactor, the blockage of the coolant flow channels, coupled with the formation of a large bubble composed of fission product gases from the damaged fuel and hydrogen from a high temperature chemical reaction between the zirconium fuel cladding and water, further complicated cooling of the reactor core. For several days engineers at the plant were forced to slowly bleed off vapor from the bubble through the pressure relief valve, until the bubble was finally eliminated and the reactor could be brought to a safe shutdown state for decontamination and repair. The reactor containment systems and the filtration system in the auxiliary building limited the actual environmental release of radioactivity during the accident to a low level. General evacuation of the surrounding population proved unnecessary.

Nevertheless, the Three Mile Island accident was quite serious. It revealed the vulnerability of complex systems such as nuclear power plants to an improbable sequence of events including both equipment and human failure. It also demonstrated that accidents can happen and probably will happen again. But Three Mile Island also confirmed the ability of the defense-in-depth design principle to protect the public from the possible consequences of nuclear plant accidents. For the Three Mile Island accident was certainly the most serious incident in the history of commercial nuclear power development, and yet nobody was injured.

But this accident also demonstrated that the principles of containment, of defense-in-depth, and conservative design are not sufficient by themselves to ensure the safety of nuclear power. It is also essential that the design, construction, and operation of nuclear power plants be very carefully monitored and regulated.

The Licensing of Nuclear Power Plants

The primary responsibility for regulating nuclear power rests with the Nuclear Regulatory Commission (NRC). This agency has the authority to issue permits to construct and licenses to operate nuclear plants. The procedure in applying for a construction permit for a nuclear power plant is outlined in table 4.[15] The application for this

TABLE 4. Steps in the Licensing of Nuclear Power Plants

1. Applicant applies for Construction Permit (CP) by submitting Preliminary Safety Analysis Report (PSAR) to Nuclear Regulatory Commission (NRC). Environmental Report submitted concurrently. Reports cover safety, environmental impact, financial capability, and antitrust implications.
2. NRC staff reviews, and on basis of review, which includes correspondence and meetings with applicant, prepares Safety Evaluation Report (SER) and Draft Environmental Statement (DES).
3. Advisory Committee on Reactor Safeguards (ACRS) reviews safety related part of application making use of PSAR and SER and holding meetings with NRC staff and applicant. ACRS prepares report to NRC.
4. Atomic Safety and Licensing Board (ASLB), using PSAR, SER, DES, and ACRS report, holds public hearings. The Board may receive comments and testimony either on the documents noted above or other issues raised by interested parties. It may also ask for any additional information deemed relevant. It recommends for or against granting a Construction Permit. Recommendation goes to NRC which can reverse ASLB. Any party in the proceeding can appeal to ASLB or NRC.
5. If application is approved, a Construction Permit is granted permitting the applicant to build plant.
6. Near the end of construction, applicant prepares and submits Final Safety Analysis Report (FSAR) along with updated Environmental Report. The above review process is repeated, with the exception that if the application is not contested, the ASLB is not required to hold public hearings.
7. If safety, environmental impact, and financial-antitrust reviews are approved, the NRC grants Operating License (OL).

permit must contain a detailed description of the plant, its site, the utility's financial and technical qualifications for constructing and operating the plant, a justification for the new plant, and two voluminous reports: a Preliminary Safety Analysis Report (PSAR) consisting of ten to twenty volumes of analysis of the safety of the proposed power plant and an Environmental Report consisting of five to ten volumes evaluating the impact of the plant on its surrounding environment. These documents are examined in great detail by the staff of the NRC. They are advised by an independent panel, the Advisory Committee on Reactor Safeguards (ACRS) composed of experts from a wide variety of technical disciplines from outside the NRC. The evaluation of the application for a construction permit will typically take about two years. During this time there will be several public hearings at which members of the public can present testimony before the Atomic Safety and Licensing Board within the NRC and may intervene by cross-examining the testimony of others. Recent legislation has been introduced in Congress that would require the NRC to provide both funding and technical support for organized intervener activities in the licensing process. Usually the evaluation process is one of iteration in that the NRC returns to the applicant with a number of questions concerning the Preliminary Safety Analysis Report and the Environmental Report, and the applicant must then respond satisfactorily to these questions, frequently by agreeing to implement changes in the proposed design or possibly changing the location of the proposed site. Occasionally the applicants have chosen to withdraw the application for a construction permit altogether at this point.[16]

If the amended application is found to be acceptable, the NRC issues a construction permit to the applicant, who then proceeds with construction of the plant. As the construction proceeds, a second major safety report, the Final Safety Analysis Report (FSAR), is prepared and submitted to the NRC. This report includes all supplements and changes made in the preliminary report and essentially documents the final design of the plant. The review process then begins once again, utilizing both the internal staff of the NRC and outside consultants, while allowing for public hearings. If the applications are approved, then an operating license is issued.

The NRC's responsibility does not end at this point, for it must maintain continual on-site inspection to ensure that the plant is operating in compliance with its operating license. As technology evolves, the NRC may require that the utility retrofit the plant to upgrade safety systems or operating procedures. Hence the safety design of the plant should not be considered fixed, but rather continually evolving as it is brought up to date with existing knowledge and experience.

The Preliminary and Final Safety Analysis Reports must not only present a detailed evaluation of the design of the nuclear plant and its various safety systems, but also carefully consider the safety aspects of the site itself. The population in the vicinity of the plant over its operating lifetime must be estimated, and the geological characteristics and seismic activity of the site, as well as the site hydrology and meteorology, carefully evaluated.

Careful consideration is given to the seismic history of the site.[17] Nuclear power plants are never knowingly sited near active geological faults. Occasionally, during ground excavation or subsequent surveys, additional faults may be discovered. If this should occur after a construction permit or even an operating license has been issued, the suitability of the site must once again be carefully examined in this new light by the NRC. Such considerations have forced the abandonment of certain sites, such as the Bodega Head and Malibu Canyon sites in California.

Nuclear power plants are designed to be able to ride out earthquakes of major intensity. That is, by a combination of structural analysis and testing during plant design, plant structures and equipment important to safety are built so that they can withstand the most severe earthquake deemed possible at the site. Since the electricity produced by such plants would be most urgently needed following a major earthquake, nuclear plants are designed so that they would not even need to be shut down during quakes of moderate intensity and could be restarted immediately after a major quake, unlike conventional plants that would probably be destroyed by such quakes. Plants located on seacoasts must be surrounded by breakwaters to protect them from possible tsunamis (tidal waves generated by earthquakes). Although there has only been limited experience with nuclear plant operation during earthquakes in this country (notably the successful operation of the San Onofre plant just north of San Diego during a large earthquake in 1971), four Japanese nuclear plants operated successfully through a severe earthquake in 1978.

A number of other safety factors must be considered in site selection. The site must be protected against possible flooding. For example, nuclear sites downstream from large dams must be protected from possible dam failures. Furthermore, the plant must be able to withstand violent storms such as tornadoes or hurricanes and the impact of debris hurled by such storms. Special consideration must be given to off-shore sites for nuclear plants constructed on mammoth barges, towed several miles out to sea, anchored there, and then surrounded by a massive breakwater.[18] The breakwater must be capable of withstanding not only large tidal waves and storms, but the direct impact of large ships as well.

The licensing and regulation procedure adopted by the NRC encourages public involvement at many stages of the licensing process. This formal provision for intervention has been greatly expanded through the courts in a host of legal actions aimed at questioning almost every aspect of nuclear power plant construction and operation. Unfortunately the present regulatory procedure for nuclear power plants is becoming so complex, ambiguous, and uncertain that it is beginning to strangle plant construction schedules. Many nuclear power plants such as the Midland plant in central Michigan and the Seabrook plant in New Hampshire have had their construction permits turned on and off again several times like a light switch in response to either administrative or legal conflicts. Both proponents and opponents of nuclear power agree that the present licensing process is a shambles. Unfortunately there does not seem to be a sufficient agreement at this time to allow for passage of legislation to remedy this dismal situation.

An Assessment of the Public Risk from Nuclear Power Plants

An awareness of the potential hazards of nuclear power has always been a major factor in power reactor development. At a very early stage the Atomic Energy Commission attempted to assess the public risk that might result from this new technology. As part of this program, the AEC in 1957 commissioned a group of physicists at Brookhaven National Laboratory to examine possible consequences of major accidents in large nuclear power plants. The report from this study, WASH-740,[19] represented not only a milestone in concern for public safety from a developing technology, but also a milestone in misquotation and misunderstanding.

In this report the Brookhaven group considered the consequences of accidents in a 200 megawatt nuclear plant located thirty miles upwind of a major city with a population of one million. They examined several accident scenarios. In the mildest case the group assumed that although all fission products in the core somehow vaporized and escaped the primary coolant system, they were contained by the containment building, and hence there was no radioactivity released to the environment. They concluded that there would be no lethal radiation exposures from such an accident.

Many nuclear critics[20] take great delight in pointing to the most serious accident examined by this group, which postulated that 50 percent of all fission products contained in the core would not only be released, which is quite impossible, but under prevailing meteorological conditions would also be dispersed over the nearby city. Needless to say the corresponding casualty estimates were large indeed, amounting to 3,400 fatalities, 43,000 injuries, and $7 billion in property damage. During the mid-1960s there was a tentative effort to upgrade the WASH-740 report to reflect the increased size of power reactors. Unfortunately this study just scaled casualty figures proportionally higher—a factor of 10 over WASH-740. Since there was no effort made to improve on the unrealistic assumptions and lack of probability estimates of WASH-740, this followup study was discontinued and no formal report was ever written.

Stimulated by increasing public concern over nuclear reactor safety and flagrant misuse of WASH-740, in 1972 the AEC commissioned an independent study under Professor Norman Rasmussen of MIT to assess as accurately as possible the public risk from commercial power reactors of the type likely to be in operation during the next several decades. The Rasmussen group examined not only the consequences of nuclear reactor accidents, but in addition they attempted to estimate the probable frequency of occurrence of these types of accidents. By using a combination of statistical and computer methods, they were able to estimate the relative risks from various types of nuclear plant accidents. Their report, referred to as WASH-1400, or the Reactor Safety Study,[21] concluded that "the risks to the public from potential accidents in nuclear power plants are very small." In particular, they concluded that

(a) The consequences of potential reactor accidents are no larger and in many cases, are much smaller than those of non-nuclear accidents. These consequences are smaller than people have been led to believe by previous studies which deliberately minimized risk estimates.

(b) The likelihood of reactor accidents is much smaller than many non-nuclear accidents having similar consequences. All non-nuclear accidents examined in this study, including fires, explosions, toxic chemical releases, dam failures, airplane crashes, earthquakes, hurricanes, and tornadoes, are much more likely to occur and can have consequences comparable to or larger than nuclear accidents.[22]

More specifically the WASH-1400 report estimates that the probable frequency of a reactor core meltdown accident is about once in twenty thousand years of operation. That is, if we achieve the projected number of four hundred nuclear plants on-line by the year 2000, then we might expect one such accident to occur at some plant in the United States every fifty years. The average consequences of each such accident were projected to be 10 deaths from acute radiation sickness plus an additional 500 deaths over several subsequent decades. Of course, the consequences of the worst possible accident are far more serious and estimated to be some 3,500 fatalities from acute radiation sickness plus 45,000 later cancer deaths and a maximum property loss of $14 billion. But the probability of such a catastrophic event is quite remote and its occurrence is estimated at once in a million years. In this regard, it should be noted that while the particular sequence of events that led to the Three Mile Island accident in 1979 was assigned a low probability by the WASH-1400 study, very similar accident sequences were examined. These studies suggested that accidents of this magnitude (causing substantial core damage but no significant radioactivity release) were quite probable in the near term.

Although these numbers sound large, they should be placed in perspective by comparing them with other risks that we routinely face in our society (see table 5). Indeed, when the risk per unit of energy produced by various sources is compared, nuclear power is seen to be a very low-risk form of energy generation, even when compared with soft energy sources such as solar or wind power (see table 6). Even the dramatic magnitude of the worst possible nuclear accident is no greater than those that might be expected from hydroelectric dam failure or explosions of liquefied natural gas storage tanks, for example.

TABLE 5. Risk of Fatality by Various Causes

Accident Type	*Total Number*	*Individual Chance per Year*
Motor vehicle	55,791	1 in 4,000
Falls	17,827	1 in 10,000
Fires	7,451	1 in 25,000
Drowning	6,181	1 in 30,000
Firearms	2,309	1 in 100,000
Air travel	1,778	1 in 100,000
Lightning	160	1 in 2,000,000
Tornadoes	91	1 in 2,500,000
All accidents	111,992	1 in 1,600
Nuclear reactor accidents[a]	0	1 in 300,000,000

Source: Reactor Safety Study, U.S. Nuclear Regulatory Commission Report WASH-1400. (Washington, D.C., 1975).

a. Since no member of the public has ever been killed by a nuclear power plant accident, the individual chance per year cannot be based on actual accident statistics, but rather must be estimated. The estimate given in this table assumes 100 operating nuclear plants.

TABLE 6. Comparison of the Risk in Man-Days Lost per Megawatt-year
Net Energy Output over Lifetime of System for Various Sources

	Occupational	Public
Coal	73	2010
Oil	18	1920
Nuclear	8.7	1.4
Natural gas	5.9	–
Ocean thermal	30	1.4
Wind	282	539
Solar		
Space heating	103	9.5
Thermal	101	510
Photovoltaic	188	511
Methanol	370	0.4

Source: H. Inhaber, *New Scientist*, May 1978, p. 444; *Science* 203: 718 (1979).

Appearing as it did at a time of greatly heightened debate over nuclear reactor safety, the WASH-1400 report naturally was immediately attacked by nuclear critics and embraced by proponents,[23] although in most cases neither group digested even a fraction of the detailed analysis outlined in its fourteen volumes. Since this reaction was anticipated, the AEC issued the report in draft form in early 1974 and invited outside groups to evaluate the study. Although various aspects of the methodology and the data base have been questioned, by and large most of these groups concluded that WASH-1400 represented a significant advance over previous attempts to estimate the risks of nuclear power.[24]

Perhaps the most serious objections have been directed at the methodology used in calculating the risk from low-probability events such as core meltdown and in estimating the consequences of reactor accidents. In the latter instance the WASH-1400 report did not use the linear radiation-dose-effect relationship commonly assumed in other studies, but introduced a dose reduction factor that lowered its estimates of radiation-induced cancers. Other aspects of the report have been questioned, such as the estimates of emergency core cooling systems reliability, the assumptions regarding effective population evacuation, and the lack of scrutability of the report.

Although there is still some disagreement about the reliability of absolute probabilities and consequences, the WASH-1400 report has come to be regarded as the most complete picture of accident probabilities associated with nuclear reactors. This study now serves as a model for risk assessment in other fields and should prove useful in providing a *relative* risk assessment comparison of alternative technologies for electric power generation.

Nuclear Accident Liability and the Price-Anderson Act

One additional issue related to safety has become a source of public confusion: insurance coverage and liability for nuclear accidents. There is a nuclear exclusion clause in most homeowner's policies, in addition to exclusions for landslides, earthquakes, and other disasters. Furthermore Congress passed in 1957, and later renewed in 1975, the Price-Anderson Act, which limits the liability of the participants in a nuclear power project in the event of a nuclear accident to $560 million. Many critics point to these facts as evidence that nuclear power is unsafe.

The reason for the nuclear exclusion clause is quite simple: if your property is damaged by a nuclear accident, the plant owners are responsible for the damage, not you. Therefore, the nuclear facility must carry liability insurance against any property loss it might cause. Your property is certainly insured against nuclear damage, but the responsibility for obtaining this insurance belongs to the utility. The Price-Anderson Act demands that nuclear power plants carry $560 million of liability coverage.

This coverage is a form of no-fault insurance in the sense that you must only demonstrate damage to your property as a result of a nuclear accident—not fault on the part of the operating utility—to receive compensation. The limit of $560 million was set for several reasons. First, there must be some limit set on the amount of liability insurance that any industry is going to be required to carry. There is no such thing as unlimited insurance coverage. Since the probability of an accident incurring damages in excess of this amount is remote (WASH-1400 estimates that such an accident would be expected to occur only once in 140,000 years of plant operation), it seems unnecessary for utilities to be required to carry additional insurance. Even if damage in excess of this coverage should occur, the spirit of the Price-Anderson Act was that the federal government would provide additional relief, as it has for other disasters such as floods.

Frequently people question why the government supplies part of this indemnity coverage (at considerable expense to the utilities in the form of premiums). The reason is again quite simple: since there has never been a nuclear plant accident, there are no statistics available of the type required by private insurance companies in underwriting various activities. Furthermore insurance companies are limited in the amount of assets they may risk in one major event. This principle applies not only to nuclear insurance, but also to fires, explosions, earthquakes, and other types of catastrophes. The $140 million liability coverage provided for each nuclear power plant by private insurance companies is the maximum amount of liability that such companies will underwrite for any single risk, whether it is an oil refinery, supertanker, aircraft, or nuclear plant.

The Price-Anderson Act was adopted originally to encourage private participation in nuclear power development. But the situation has changed since 1957. Today almost two hundred nuclear plants are either in operation or under construction. Insurance companies are now willing to provide much of the liability insurance ($140 million) for a nuclear plant, while an additional amount ($340 million) is obtained from a $5 million premium assessed retroactively to each nuclear plant.[25] The limit on liability is still retained since unlimited liability would cause licensing problems (utilities or vendors would be challenged about their ability to cover an improbable but potentially exorbitant set of claims) and smaller contractors would be discouraged from participating in nuclear activities.

The United States Supreme Court unanimously upheld the constitutionality of the Price-Anderson Act in 1978, noting that the act "provides a reasonably just substitute for the uncertain recovery of damages of this magnitude from a utility or component manufacturer whose resources might well be exhausted at an early stage."[26] The Court furthermore noted that the law also "eliminates the burden of delay and uncertainty which would follow from the need to litigate the question of liability after an accident."[27]

Environmental Impact of Nuclear Power Plants

The major environmental impact of any power plant is caused by the various types of discharges that such plants release to the environment. We need only glance at the stacks towering above a fossil-fueled power plant to realize that substantial quantities of combustion products in both gaseous and solid forms are discharged directly to the atmosphere. Although nuclear plants do not release combustion products to the environment and therefore are far superior to fossil-fueled units from this perspective, they do release minute quantities of radioactive materials. Furthermore all large power plants discharge significant quantities of waste heat to the environment, either directly into adjacent bodies of water or into the atmosphere. We will examine each of these factors in assessing the environmental impact of nuclear power plants.

Thermal Discharges from Power Plants
Any electric power plant based on a thermal cycle will reject some 60 to 70 percent of the thermal energy it produces directly into the environment as waste heat. This unfortunate feature is not due to

sloppy engineering, but rather to fundamental limitations of the basic laws of physics (thermodynamics). A 1,000 megawatt-electric nuclear plant will typically discharge 2,000 megawatts of waste heat into the environment. A comparable fossil-fueled plant will discharge 1,500 megawatts of heat because of its slightly higher thermal efficiency (40 percent as compared with 33 percent for a nuclear plant). Therefore, while thermal discharge problems are not unique to nuclear power plants, they are somewhat more significant in this type of generating unit.

In older power plants, once-through cooling cycles were used in which condenser cooling water was drawn directly from an adjacent lake or river, passed through the condenser, and then discharged at somewhat higher temperatures back into the body of water. It rapidly became evident that such discharges could significantly modify the local aquatic environment.[28] Therefore most modern nuclear plants are designed with closed cooling cycles, usually based on natural draft cooling towers that discharge waste heat directly into the atmosphere.[29] However, even these closed cycles can have significant environmental impact. For example, the makeup water requirements for a 1,000 megawatt plant to replenish water evaporated in the cooling towers will be some 50,000 gallons per minute. This must be drawn as a permanent drain from a nearby body of water. Moreover, the thermal discharge from these enormous plants directly into the atmosphere can have effects on the local meteorology, causing fogs or icing. There is some concern that chemicals such as chlorine that are used to treat condenser cooling water may find their way into the environment with harmful effects. Many ecologists now regard the relatively small temperature increases associated with once-through cooling as preferable to the closed cycles based on cooling towers.

Liquid, Solid, and Gaseous Wastes
Fossil-fueled electric generating units discharge enormous quantities of combustion products directly into the atmosphere. The wastes discharged by a modern 1,000 megawatt coal plant are listed in table 7.[30] It is interesting, and somewhat alarming, to note that some 24,000 tons of SO_2 are discharged each year from such a coal-fired plant, which on contact with water vapor can be converted into sulfuric acid and can significantly damage the environment. Nitrous oxide can seriously damage lung tissue. Carbon monoxide is another pollutant from fossil-fueled plants. Although the 6 million tons of CO_2 discharged from such a coal-fired plant each year are not a direct health hazard, the large quantities of CO_2 generated by fossil-

TABLE 7.   Waste Material from a 1,000 Megawatt-electric Power Plant

	Coal-Fired	Nuclear
Thermal efficiency (%)	39	32
Thermal wastes (in megawatts)	1570	2120
Solid wastes		
Fly ash or slag (tons/year)	330,000	0
(cubic feet/year)	7,350,000	0
(railroad cars/year)	3,300	0
Radioactive wastes		
Spent fuel (assemblies/year)	0	160
(railroad cars/year)	0	5
Gaseous and liquid wastes (tons/year)		
Flyash (particulates)	2,000	0
Sulfur dioxide	24,000	0
Carbon dioxide	6,000,000	0
Carbon monoxide	700	0
Nitrogen oxides (as NO_2)	20,000	0
Mercury	5	0
Arsenic	5	0
Lead	0.2	0
Radioactive gases or liquids		
(maximum whole body dose at		
plant boundary in mrem/year)	1.9	1.8

fueled plants could cause an increase in the atmospheric concentration of CO_2, leading to a rise in average global temperatures (the "greenhouse" effect). Furthermore significant quantities of toxic materials such as arsenic, lead, mercury, and even radioactive materials such as radon are released up the stacks of coal plants.

Burning coal typically produces between 10 to 20 percent ash, which amounts to some 330,000 tons of flyash per year that must be removed and disposed of in some suitable manner. Engineers have not been particularly successful at finding a use for this enormous quantity of waste material, and it is typically trucked off and used as landfill.

Although nuclear power plants do not release waste to the environment in an uncontrolled fashion, they do produce significant quantities of radioactive materials or radioactive "waste" during the fission chain reaction in the form of spent fuel that must be carefully removed from the plant and eventually deposited in some suitable disposal facility. Although the mass of such waste is typically quite small (several tons per year per plant), it nevertheless requires great care in handling and treatment. We will consider the topic of radioactive waste disposal in detail in chapter 4.

Radioactivity Releases from Nuclear Power Plants

It is inevitable that small quantities of radioactive material will be released from nuclear power plants into the environment. It is impossible to have zero release of radioactive materials from such plants, just as it is impossible to achieve zero release of pollutants from fossil-fueled plants—or any other process for that matter. However, nuclear power plants are designed so that these releases are kept far below those levels that might have significant effects on public health and at only a tiny fraction of the natural radioactivity level in the environment. Present federal regulations[31] require that radioactivity releases be reduced *as low as reasonably achievable* (ALARA) with present technology. The emission levels from nuclear plants are now so low that it is extremely difficult to measure releases and verify that the plant is complying with federal regulations.

Put another way, radioactivity release standards are set several orders of magnitude below any clinically observable level, compared with stack emission standards for fossil-fueled plants, which are close to toxic levels (see fig. 16).[32] For example, a tenfold increase in SO_2 emissions above present standards would almost certainly lead to a noticeable increase in hospital admissions and morbidity rates. In contrast, if a nuclear plant were to release ten or even one hundred times the present standard, there would still be no clinically observable effects. This discrepancy is due in part to the ease with

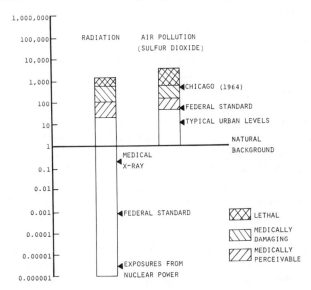

Fig. 16. A comparison of standards governing radiation and air quality.

which very low radioactivity levels can be detected. For radioactive pollutants the problem of detecting concentrations of 1 part per billion or even 1 part per quadrillion is almost routine and continuous monitoring devices are not too difficult to obtain. For chemical pollutants the detection of concentrations of 1 part per million is often difficult, and continuous monitors for chemical pollutants are essentially nonexistent.

The principal sources of radioactive material that might be released from the plant during normal operation include (1) radioactive fission products (primarily gases) produced in the fuel that leak out of the fuel elements through tiny cracks in the cladding, into the coolant of the reactor system, and subsequently past valves, fittings, packing, and other mechanical sealing devices into the containment building, (2) radioactive material in the coolant of boiling water reactors that can be carried into the turbine building, where it may again leak past mechanical sealing devices, and (3) a number of radioactive materials produced by neutron activation of impurities in the coolant water or by slow corrosion of components in the reactor core that may be dispersed through the primary coolant system.

Most of the solid radioactive waste produced in the reactor core remains entrained in the fuel pellets and does not escape from the fuel elements. However, when the reactor is refueled, the spent fuel elements containing these solid wastes must be withdrawn and reprocessed, and the radioactive waste disposed of in some suitable fashion. Radioactivity releases from these fuel reprocessing operations will be discussed in chapter 4.

The minute quantitites of radioactive materials that escape from the fuel elements into the coolant take both gaseous and liquid forms. The gaseous radioactive material is usually collected by filters, compressed, and held in storage tanks until a large fraction of its radioactivity decays away. At some later time, when its radioactivity is sufficiently low, it is released through the building ventilation system in a controlled manner and dispersed into the atmosphere. The primary radioactive gases of concern are krypton-85, iodine-131, tritium, and carbon-14 (as CO_2 gas).

In a similar manner all liquid radioactive materials are separated out by filtration systems and held to allow decay before being dispersed into large bodies of water. Federal regulations restrict both the amount of radioactivity and the concentration at which it may be released into the environment. They also dictate the amount of processing of the radioactive effluent that is required, the time it must be held, and the rate at which it is discharged as a plant effluent.[33] Presently the release levels of liquid radioactive wastes are restricted so that the exposure of an individual in an unrestricted area due to discharge liquids is no greater than 5 mrem per year. (Here *mrem* stands for milli-Roentgen-equivalent-man, the standard unit for measuring radiation dose.)

Gaseous releases are restricted so that mythical individuals sitting on the site boundary fence for twenty-four hours a day, 365 days a year, will be exposed to less than 5 mrem whole body dose or 15 mrem to the thyroid. The average neighbor of the plant can receive no more than 1 mrem per year. To place these exposures in perspective, the natural background radiation exposures are typically between 80 and 150 mrem per year (with fluctuations of about 20 to 30 mrem) and medical exposures received by the public amount to some 70 to 80 mrem per year (see table 8). Therefore the radioactivity released from nuclear plants is restricted so that the increased exposure to an individual in the vicinity of the plant is less than 1 percent of the exposure he would receive from natural sources (see table 9). The actual exposures from nuclear plants are less than the fluctuation in this natural exposure. Members of the

TABLE 8. Average Environmental Radiation Doses to the
General Population

Source	Dose (mrem/year)
Natural radiation[a]	
Cosmic radiation	28
Cosmogenic radionuclides	0.7
External terrestrial	26
Radionuclides in the body	24
Rounded total	80
Man-made radiation[b]	
Medical and dental X-rays	50-100
Fallout (average from 1951-76)	7
Luminous watches	
Radium	3
Tritium	0.5
Airplane travel	0.3 mrem/hour
Color television	negligible
All consumer products	less than 1 mrem/year
Nuclear power (400 gigawatts)[c]	0.05
Rounded total	50-100
Total background dose	150-200 mrem/year

a. National Council on Radiation Protection Report no. 45 (1975), p. 108.

b. Report of United Nations Scientific Committee on the Effects of Atomic Radiation (1977).

c. B. L. Cohen, *Am. Sci.* 64: 550 (1976).

TABLE 9. Various Radiation Protection Standards Based on Maximum
Permissible Dose Rate to Thyroid from Iodine-131

Type of Person	Source of I-131	Legal Reference	Maximum Permissible Dose Rate (in mrem per year)
Occupational worker	Occupational	10 CFR 20	30,000
General public	Medical	10 CFR 20	unlimited
General public maximum exposure	All but nuclear power	10 CFR 20	500
General public average exposure	All but nuclear power	10 CFR 20	170
General public	Nuclear power	10 CFR 50, appendix I	15

public living some distance from the plant would receive far lower
exposures. Recent studies[34] have indicated that the radiation doses
from the airborne radioactivity released by coal plants due to trace
concentrations of uranium and thorium in coal exceed those of
nuclear plants. The average public exposure from nuclear power
plants in the year 2000, assuming some four hundred plants are
on-line at that time, is estimated to be about 0.05 mrem per year—
1/3000 of that received from natural sources of radiation in our
environment.[35]

Therefore any possible effects on public health from nuclear
plant emissions would be overwhelmed by those resulting from
natural sources of radioactivity or other contaminants in our environment and are therefore essentially impossible to detect.
Radiological health physicists are concerned that the exceptionally
low federal standards set for radiation emissions from nuclear power
plants are illogical and set a dangerous precedent for future environmental standards.[36] The essential philosophy that has been
adopted by federal regulators is that, if the technology is available to
reduce radioactivity releases to these incredibly low levels, then the
nuclear plants must be designed to operate routinely below these
levels regardless of the cost or difficulty in monitoring the compliance with these limits. Such standards almost totally ignore the
fact that there is no evidence whatsoever indicating that operating
the plant at ten times or even one hundred times larger releases
would cause any detectable health effect. Furthermore the limits

applied to nuclear power plants are almost a factor of ten more stringent that those applied to medical or industrial applications of radioactive materials. But since the nuclear power industry has developed (at enormous expense) the capability of pushing releases down to these low levels, they are now required by law to operate below these levels. Perhaps after some experience has been gained in operating at these levels, the limits on radioactive emissions will be reduced even further (although there is no reason from the public safety standpoint for doing this), corresponding to larger and larger investments in radioactive materials treatment equipment in nuclear plants.

Economics

We have noted that the choice among various forms of energy production will be dictated by economics, since most other factors involved in such comparisons, such as resource availability, environmental impact, and public safety, eventually will be measured in terms of cost. There is little doubt that nuclear power plants are generating electricity at a lower cost than fossil-fueled plants. Data for the past several years[37] (see table 10) indicate that electric generating costs for nuclear plants are some 20 percent below those of coal-fired units and almost a factor of 3 cheaper than oil-fired plants.

The economic advantages presently exhibited by nuclear plants are due in part to the rapid escalation in fossil-fuel costs (roughly a factor of 4) since the OPEC oil boycott in 1973. But uranium prices have also skyrocketed, from the $8 a pound for U_3O_8 in the early 1970s to $40 a pound. However, since only about 15 percent of the cost of nuclear power generation is attributable to fuel charges and only about 5 percent to uranium ore prices, nuclear power has been able to retain its economic attractiveness in the face of rising fuel costs. For example, a $10 a pound increase in uranium prices raises electric generating costs by only about 1 mill per kilowatt-hour (out of a present total of 15 mills per kilowatt-hour); a $10 per separative work unit increase in enrichment costs corresponds to only a 0.25 mill per kilowatt-hour increase.[38]

This relative insensitivity of nuclear power costs to fuel charges is countered by the highly capital-intensive nature of this form of power generation. Over 80 percent of the electric generating costs of nuclear units can be attributed to the capital costs of the plant (including interest charges). These costs have been soaring at an alarming rate. Nuclear plants ordered in 1970 were based on pro-

**TABLE 10. Comparative Economic Data for Alternative Electric
Generating Units[a]**

Electric Generating Costs (in mills per kilowatt-hour)		
Plant Type	*1978*	*1985*
Oil	40.0	55.1
Coal	23.0	42.0
Nuclear	15.0	38.3

A Breakdown of Electric Generating Costs Projected for 1985			
	Oil	*Coal*	*Nuclear*
Capital cost	13.7	20.8	25.2
Fuel cost	39.7	17.0	10.0
Operating and maintenance cost	1.7	4.2	3.1
Total	55.1	42.0	38.3

Reliability of Electric Power Plants for 1978 (in percentage)			
	Oil	*Coal*	*Nuclear*
Availability factor	77.2	77.8	76.6
Capacity factor	50.7	55.1	68.0
Forced outage factor	9.0	12.6	10.5

a. Example: The average homeowner presently consumes roughly 500 kilowatt-hours
of electricity per month. Hence, at a generating cost of 15 mills per kilowatt-hour,
the monthly charge for this electricity would amount to $7.50. When transmission
and service charges are included, the bill increases to $20 to $30.

jected capital costs of $300 per kilowatt of capacity ($300 million for
a 1,000 megawatt plant); plants ordered today will cost more than
$1,000 per kilowatt capacity when interest and inflation are included.

This dramatic escalation in capital costs has prompted a serious
reevaluation of the relative economic attractiveness of nuclear
power generation. It should be recognized that the present economic
advantages exhibited by nuclear power are based on operating
plants that were ordered in the late 1960s when capital costs were
significantly lower. The projections that nuclear plants ordered
today would benefit from similar cost advantages is due in large part
to the rapid escalation in the construction costs of fossil-fueled
plants. For example, recent bids[39] for a 2,400 megawatt power plant
for upstate New York recently estimated capital costs for a nuclear
plant at $1,379 per kilowatt capacity compared with $1,337 per
kilowatt capacity for a coal-fired plant (equipped with stack gas
scrubber units)—only about 3 percent less expensive. The capital

costs of modern coal-fired and nuclear units are now almost identical in many parts of the country.

The escalating costs of generating plants have stimulated a number of plant order deferrals and cancellations by the electric power industry. Although the slowdown in plant construction was also due in part to the economic recession of the mid-1970s and the difficulty in obtaining the enormous capital required for power plant construction, it has hit nuclear power generation somewhat harder than fossil-fueled units.[40] This is due to several factors, including the temporary leveling off in the growth of electric power demand that was stimulated by this recession and the significant increase in the time required for nuclear plant construction. The construction delays are due in part to the complexity (and confusion) of regulatory requirements, the extensive public hearings and evaluation processes that encourage active intervention, particularly litigation, by groups opposed to nuclear plant construction, and by the lack of a coordinated federal government policy regarding nuclear power. (Interestingly enough, the ever-growing mass of EPA regulations affecting coal-fired plants has dramatically increased their construction time as well, to the point where it now approaches that of nuclear plants.) When it is recognized that almost 60 percent of the capital cost of a plant is due to interest and escalation charges, there is little wonder that there is a strong impetus to remove such artificial barriers to plant construction.

A variety of proposals has been offered to accelerate the plant construction process. For example, there are presently bills before Congress that would speed up the plant licensing process by allowing a utility to have a number of potential power plant sites approved far ahead of the time when actual orders are placed for the plant. Furthermore the acceptance of preapproved, standardized plant designs would significantly shorten the regulatory review process. Perhaps the most controversial suggestion is to somehow limit the degree to which groups opposed to nuclear power can delay plant construction through intervention, both legal and illegal. It is rather ironic that a major criticism leveled at nuclear power by such groups concerns its rapidly rising capital costs, which these groups have stimulated by delaying plant construction. For example, delays in construction in the Seabrook, New Hampshire, plant have raised the cost of this plant by an estimated $419 million.[41] The seven-year construction delays in a two-unit plant at Midland, Michigan, which were not entirely due to intervention,[42] have raised the projected cost of the plant from $300 million to $1.7 billion.

Unless a significant reduction in construction time can be

achieved and some mechanisms found to help electric utilities raise the enormous capital required for construction of generating plants, both nuclear and coal-fired, the future may look rather bleak for all forms of electric power generation. The slowdown in nuclear plant orders over the past few years has seriously affected the nuclear power industry, which had built up a capability for handling some twenty to thirty new plant orders each year. There is some evidence that this industry is beginning to disintegrate. Engineers are beginning to leave the major nuclear equipment vendors since these companies can operate on backlog and fuel resupply orders only for a short period. Many smaller nuclear equipment suppliers are also beginning to suffer from the sales decline.

Even more serious is the possibility that by the mid-1980s the electric utilities in this country will be facing a significant shortage in reserve capacity unless steps are quickly taken to put power plant construction back on schedule. This will lead to an inevitable decline in system reliability in the form of brownouts and eventually blackouts during periods of peak electrical demand.[43]

In examining the economics of nuclear power we have focused on monetary concerns, but we should examine as well the economics of energy investment in this technology. It is sometimes conjectured that this energy investment may be so large that it can never be repaid by actual plant operation. Table 11 gives a detailed comparison of the energy efficiency of nuclear and fossil-fueled power plants. Numerous studies of the energy efficiency of alternative production facilities have indicated that it takes roughly one year for a nuclear plant to pay back the energy invested in its construction and operation.[44] Most of the energy investment occurs in the enrichment process of the nuclear fuel preparation. The energy investment in plant construction is paid back rapidly, in several months of plant operation. These estimates indicate that nuclear power generation is a highly energy-efficient process.

TABLE 11. Energy Analysis of Electric Power Plants[a]

| | Nuclear Plants | | Fossil-Fueled Plants | |
	Light Water Reactor	Liquid-Metal-Cooled Fast Breeder Reactor	Coal	Oil
Net energy ouput gigawatts/year)	24	24	24	24
Energy released by power plant fuel	75	61.5	65.4	61.5
Indirect energy requirements				
Transportation	0.01	0.01	0.38	2.07
Mining and milling	0.12	0	–	–
Conversion	0.03	0	–	–
Enrichment	2.15	0	–	–
Fuel fabrication	0.03	0.03	–	–
Fuel reprocessing	0.07	0.05	–	–
Mining and cleaning	–	–	0.43	–
Extraction	–	–	–	0.47
Sulfur removal	–	–	–	0.75
Plant construction	0.47	0.59	0.37	0.37
Other construction	0.08	0.02	0.04	–
Direct energy requirements				
In-plant processing	–	–	1.07	0.46
Pollution control	–	–	3.24	–
Plant pumps, etc.	3.85	4.47	2.29	2.29
Total energy requirements	6.81	5.17	7.82	6.41

Source: R. H. Fischer and R. S. Palmer, "The Energy Efficiency of Electric Power Plants" (Paper presented at the Sixteenth Annual ASME Symposium on Energy Alternatives, Albuquerque, N. Mex., 1976).

a. Based on 1,000 megawatt plants with 80 percent capacity factor, thirty-year plant life, and a total of 24 gigawatt-year net energy output.

4

The Nuclear Fuel Cycle

The safety and environmental impact of nuclear power plants have become the subjects of increased public concern during the past decade. But generating plants are only one aspect of nuclear power. More recently public attention has encompassed the entire *nuclear fuel cycle*—the operations involved in the preparation, utilization, reprocessing, and disposing of nuclear fuels.[1]

Nuclear fuels are totally different from fossil fuels in several respects. Nuclear fuel material such as uranium must undergo a number of sophisticated and expensive processing operations before it is inserted into the reactor core. It is then "burned" in the reactor for several years before being removed. Even after several years of use, the fuel possesses a significant concentration of fissile material. Therefore, after the spent fuel is removed from the reactor core, it can be chemically reprocessed to extract the unused fissile material, which can then be refabricated into new fuel elements. The by-product waste from the reprocessed fuel includes highly radioactive fission products, and its disposal requires great care.

The safety and environmental impact of each nuclear fuel cycle activity must be examined carefully in any consideration of the future role of nuclear power. Moreover, since the primary economic advantages exhibited by nuclear plants are a consequence of their extremely low fuel costs, the economics of nuclear fuel preparation, reprocessing, and disposal must also be considered.

An Overview of the Nuclear Fuel Cycle

The various stages of the nuclear fuel cycle are illustrated in figure 17.[2] These include mining, milling, enrichment, fuel fabrication, fuel burnup, spent fuel storage and decay, spent fuel reprocessing, and radioactive waste disposal.

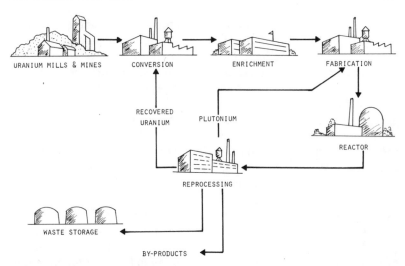

URANIUM MILLS & MINES   CONVERSION   ENRICHMENT   FABRICATION

RECOVERED URANIUM   PLUTONIUM

REACTOR

REPROCESSING

WASTE STORAGE

BY-PRODUCTS

Fig. 17. The nuclear fuel cycle. (*Courtesy of the United States Department of Energy.*)

Mining. Most of the uranium ore mined in the United States comes from sandstone deposits in the Colorado Plateau and the Wyoming Basin. Both underground and open-pit techniques are used in a manner similar to that used in other low-grade ore mining. There is considerable uncertainty about the extent of our domestic uranium ore resources. Present United States Department of Energy estimates place proved plus probable plus possible resources of U_3O_8 at less than \$50 per pound (forward costs) at $890,000 + 1,395,000 + 2,080,000 = 4,365,000$ tons.[3] These estimates suggest that there should be no difficulty in fulfilling lifetime ore commitments for light water reactors through the year 2000 for installed capacities ranging from 300 to 500 gigawatts. Beyond this point, further expansion of light water reactor capacity becomes problematic.

Milling. Milling is necessary to extract and concentrate uranium from the raw ore. The ore is first pulverized, and then a solvent extraction process is used to produce *yellowcake*, a crude oxide containing some 70 to 90 percent U_3O_8.

Enrichment. Essentially all power reactors (with the exception of the Canadian heavy water reactors or the early British gas-cooled, graphite-moderated reactors) utilize enriched uranium, that is, uranium with higher than the natural 0.7 percent concentration of U-235. The enrichment of uranium is a difficult and expensive process since it involves separating two isotopes, U-235 and U-238, that have little mass difference and essentially no chemical difference. A variety of techniques have been used or proposed, including electromagnetic separation, gaseous diffusion, gas centrifuge or nozzle separation, and laser isotope separation. At the present time gaseous diffusion continues to be the most common method of uranium enrichment, although both gas centrifuge and nozzle separation plants are under construction, and laser methods hold promise for the future. Enrichment likely will remain an expensive process; it presently accounts for some 30 percent of nuclear fuel costs and some 75 percent of energy investment in nuclear power generation.

Fuel Fabrication.   Enriched uranium is chemically converted into a ceramic powder such as UO_2 or UC and compacted into small pellets. These pellets are loaded and sealed into metal tubes to produce reactor fuel elements, which are then fastened together into bundles known as fuel assemblies.

Fuel Burnup in the Reactor Core.   The fuel assemblies are loaded into the reactor core for power production and are typically irradiated for several years. The fuel lifetime may be limited by criticality, that is, the fissile concentration may drop too low to sustain the chain reaction, or by radiation damage sustained by the fuel elements during operation.

Spent Fuel Storage and Decay.   After being irradiated in the reactor core, the fuel is intensely radioactive with fission products and other radioactive nuclei due to neutron absorption. The spent fuel is removed from the core and stored in water pools in the plant for several months to allow the short-lived radioactive nuclei to decay away. It is then loaded into heavily shielded and cooled casks for shipping to reprocessing or storage facilities by either truck or rail. The shipping containers are carefully designed to ensure their integrity in the event of any conceivable shipping accident.

Spent Fuel Reprocessing.   The spent fuel discharged from a nuclear power reactor contains a significant quantity of fissile material. For example, each kilogram of spent light water reactor fuel contains roughly 8 grams of unused uranium-235 and 6 grams of fissile plutonium.[4] This material can be extracted and reloaded into fresh fuel elements. The principal scheme for commercial recovery of uranium and plutonium from low-enrichment light water reactor fuels is the Purex process.[5] This method has been used on a commercial basis throughout the world for almost two decades. However, the plutonium separated from spent reactor fuel is a possible source of material for nuclear weapons. Therefore the United States government has deferred indefinitely the reprocessing of spent commercial power reactor fuel. A number of other nations are either operating or constructing spent fuel reprocessing facilities and have every intention of proceeding with plutonium recycling in light water reactors.

Radioactive Waste Disposal. Most public attention concerning the nuclear fuel cycle has been directed toward the disposal of high-level radioactive waste produced by nuclear power reactors. Most of the radioactivity produced in power reactors will decay away quite rapidly following reactor shutdown and removal of spent fuel elements from the reactor core. However, a significant fraction of the high-level radioactivity induced in the fuel is due to fission products and actinides that will remain toxic for thousands of years. To date there has been no commitment to a specific scheme for permanent disposal of these wastes. In the event that the present moratorium on spent fuel reprocessing continues (a "stowaway" fuel cycle), the radioactive spent fuel assemblies will be placed indefinitely in a retrievable storage facility. The most attractive proposal for handling high-level waste involves converting it to a stable form such as glass or cement and then encapsulating it in stainless steel canisters. The waste would then be shipped to a federal waste repository for permanent disposal. This would involve deep burial beneath the earth in rock formations characterized by exceptional geologic stability.

This sequence of operations characterizes the uranium-plutonium fuel cycle used for light water reactors. Even this cycle has a number of possible options: (1) a "throwaway" cycle in which spent fuel is treated as waste for permanent disposal, (2) a "stowaway" fuel cycle in which spent fuel is stored in a manner that does not preclude the later recovery and utilization of its fissile energy content, (3) reprocessing to recover uranium only, and (4) reprocessing to recover both uranium and plutonium. The choice among these various options is not dictated by ore resources and economics alone, but by political considerations as well because of the potential impact of the nuclear fuel cycle on nuclear weapons proliferation. We will consider this topic in detail later in this chapter and in chapter 5.

A variety of other fuel cycles can be employed in different reactor types.[6] For example, thorium/U-233 fuel cycles can be used in light water, heavy water, or high-temperature gas-cooled reactors, although in all cases fissile material such as U-235 or plutonium must be added to thorium to achieve criticality. The greatest potential for resource utilization is provided by breeder reactors such as the liquid-metal-cooled fast breeder reactor and the gas-cooled fast breeder reactor. In these reactor types some 30 to 50 percent more fissile material is produced by neutron transmutation of uranium-238 into plutonium than is consumed during power production. In this way the breeder reactor can utilize some 60 to 70 percent of the energy available in natural uranium compared to the roughly 1 to 2 percent used by light water reactors. The more efficient utilization of available uranium fuel resources is the primary justification for such breeder reactors. These reactors can be fueled either with natural uranium (or thorium) or depleted uranium from the tails produced by enrichment plants.

The fuel requirements of various reactor types and fuel cycles have been compared in table 12.[7] It should be recognized that there will be a substantial interaction among these various fuel cycles in a mature nuclear power industry. The substantial amounts of plutonium produced by light water reactors can be directly recycled or can be used to fuel the first generation of fast breeder reactors. Similarly the excess plutonium produced by breeder reactors can be

TABLE 12.   Lifetime Uranium Requirements for Various Reactor Types and Fuel Cycles[a]

Reactor Type	Fuel Cycle	Uranium Requirements (in short tons)
Light water reactor	U, no recycling	6,410
	U, U recycling	5,280
	U and Pu recycling	4,340
	Th + U, U recycling	3,650
Heavy water reactor	Natural U, no recycling	5,263
	Natural U, Pu recycling	2,861
	Pu-Th, U recycling	2,210
High-temperature gas-cooled-reactor	U-235 + Th, U recycling	2,970
Liquid-metal-cooled fast breeder reactor	U + Pu, recycling	60

Source: T. H. Pigford, *IEC Fund.* 16:61 (1977).

a. Based on a 1,000 megawatt-electric plant operating for thirty years at 80 percent capacity factor.

used to fuel still more breeder reactors or fed back into the light water reactor fuel stream.

A number of issues involving the nuclear fuel cycle have surfaced in the debate over nuclear power.[8] The primary safety concerns arise because of the presence of radioactive materials at various stages of the fuel cycle. Although one most commonly thinks of the large inventories of radioactivity in spent fuel and radioactive waste, there are also radiological concerns in the mining of uranium ore and the processing of nuclear fuels. Particular attention has been directed at the production, extraction, and handling of plutonium in the fuel cycle. Not only is this substance extremely toxic, but it is also a prime ingredient in the fabrication of nuclear weapons. Its presence in the fuel cycle raises questions concerning the possibility of clandestine diversion of plutonium on either a subnational or international level. There are also somewhat less dramatic concerns about the extent of uranium or thorium resources, as well as the expense of the elaborate operations involved in nuclear fuel production and reprocessing and the energy investment in these operations.

Uranium Mining, Milling, and Enrichment

The mining and processing of uranium ore into nuclear fuel involves a number of complex and expensive operations. Fortunately the expense of these operations is far outweighed by the incredible energy concentration of this fuel. The consumption of one gram of uranium-235 produces 7,300 kilowatt-hours of electric energy, the equivalent of 13 barrels of oil or 3 tons of coal. Therefore, even though nuclear fuel is rather expensive on a per unit mass basis (a typical nuclear fuel assembly is worth several hundred thousand dollars), it still contributes less than 15 percent of the cost of nuclear-generated electricity. In fact, on a per unit of energy basis, nuclear fuels cost only one half as much as coal and one fourth as much as oil.

We have already noted the uncertainty in the extent of our domestic uranium and thorium resources. In this section we will discuss other aspects of the "head end" of the nuclear fuel cycle, the operations involved in mining, milling, processing, enriching, and fabricating nuclear fuel.

Uranium Mill Tailings
Much of the technology involved in the exploration and production of uranium has been adopted from other mineral industries. However, uranium ore has one characteristic not common to other com-

mon minerals: radioactivity. This characteristic has created the possibility of novel approaches to the exploration for uranium, but it has also created new problems for mining and processing. Of particular concern have been possible hazards arising from the residual radioactivity released by the uranium tailings that result from the milling process.[9] The tailings residue from the uranium ore milling is typically stockpiled adjacent to the milling facilities. These uranium tailings will release small quantities of a radioactive gas, radon, to the atmosphere. (Comparable quantities of radon are also released from open-pit mining activities.) During the early days of the uranium mining industry, these tailings, which have the consistency of fine sand, were carted off by local contractors and used in mixing concrete for building foundations, most notably in the Grand Junction and Durango, Colorado, areas. It was discovered in the 1960s that the radon level in buildings constructed on such foundations exceeded radiological health standards. Therefore this practice was discontinued, and the foundations in question were replaced.

There has been continuing concern, however, that the radon escaping from the tailings piles themselves could have an adverse effect on public health. In fact some estimates suggest that if these tailings are not adequately handled, they may eventually represent the dominant contribution to radiation exposure in the nuclear fuel cycle.[10] Although the radon level at more than one kilometer from tailings piles is indistinguishable from background levels of this nuclide,[11] local concentrations may become quite high. Therefore, to control the release of this radioactive material, the NRC is developing plans[12] to require the uranium milling industry to cover or pave over existing tailings piles to reduce radon emission to natural background levels. There have also been proposals that future milling operations be required to bury tailings, perhaps in the original mining pits.

Enrichment

An essential step in the nuclear fuel cycle is the enrichment of uranium from its natural concentration of 0.7 percent U-235 to the higher concentrations required in modern power reactor fuels (for example, 2 to 3 percent for light water reactors, 93 percent for high-temperature gas-cooled reactors). To achieve the enrichment of natural uranium, one must first convert the uranium compound to a gaseous form, uranium hexaflouride, UF_6. In the two principal processes in use today, gaseous diffusion and gas centrifuge enrichment, this separation is achieved by utilizing the small mass difference between the two isotopic forms of gaseous UF_6 to separate the uranium isotopes. Since this mass difference is small, the amount of separation that can be achieved in a given device is similarly quite small. Hence, to achieve appreciable separation, a large number of separation devices must be arranged in a series or cascade. It requires a massive investment of both capital and energy to enrich uranium, and in this country this process has been under the exclusive control of the federal government.

In this section we will examine several methods for uranium enrichment: gaseous diffusion, gas centrifuges and nozzle devices, and laser photoexcitation. A fourth technique, electromagnetic separation, was used during the early days of the Manhattan Project but was quickly abandoned because of its enormous expense. The principal technique in use today is still gaseous diffusion, although it is expected that the gas centrifuge method will eventually prove superior.

Gaseous Diffusion. The gaseous diffusion separation method utilizes the difference in the rates at which gases of different molecular weights diffuse through a porous membrane or barrier to separate the isotopic forms of UF_6 gas. The rate of diffusion is inversely proportional to the square root of the molecular weight of the gas. Therefore the best separation that can be achieved by diffusion through a single membrane, that is, the separation factor for the process, is limited by the square root of the mass ratio of the isotopic forms of UF_6 to 1.0046. Since the enrichment per stage is very small, a large number of stages in series or cascade is required to produce significant enrichment. For example, the production of 3 percent enrichment from natural uranium feed requires about fifteen hundred stages in cascade.

The effort involved in separating isotopes is referred to as *separative work* and is measured in *separative work units* (SWUs). A quantitative definition of an SWU requires more effort than is warranted here,[13] so suffice it to note that the production of 1 kilogram of 3 percent enriched uranium from 4 kilograms of natural uranium feed requires 1 kilogram SWU.

The basic separation stage in a gaseous diffusion plant consists of the porous barriers, a compressor to maintain a pressure differential across the barriers, and a heat exchanger to remove the heat of compression. High-pressure UF_6 is fed into tubes made of the barrier material. The UF_6 diffusing through these barriers is slightly enriched due to the difference in isotopic diffusion rates. This slightly enriched UF_6 is drawn off at lower pressure, while the remaining material, which is now slightly depleted of U-235, is drawn off from the end of the stage. Large numbers of stages are coupled together into a two-stream cascade in which one stream becomes progressively enriched while the other becomes depleted.

The gaseous diffusion stages require large amounts of power to maintain the pressure differential across the thousands of barriers. For example, the three gaseous diffusion plants operated by the Department of Energy at Oak Ridge, Tennessee, Paducah, Kentucky, and Portsmouth, Ohio, require almost 6,000 megawatts of power if run at full capacity. The energy required for uranium enrichment using the gaseous diffusion process comprises almost 75 percent of the total energy investment in nuclear power generation. However, this energy investment is still less than 9 percent of the energy that will be produced by the fissioning of the enriched uranium in a nuclear power reactor, including the energy lost in converting fission heat energy to electricity through a steam thermal cycle.[14] (Refer back to table 11 on page 93 for a detailed breakdown of the energy investment in nuclear power generation.)

Gas Centrifuge Techniques. The idea of separating the two isotopic forms of UF_6 gas using high-speed centrifuges was considered in the early days of the Manhattan Project. However, this approach was abandoned because of the difficulty in achieving the high rotation speeds (50,000 rpm) necessary for separation without destroying the centrifuges. Developments in centrifuge design during the early 1960s have led to a renewed interest in this method.[15]

Since centrifuge stages are characterized by much larger separation factors (1.1 to 1.4), only a relatively few centrifuges need be connected in series to achieve substantial enrichment. However, since the flow rates possible in centrifuges are much lower than in gaseous diffusion stages, large numbers of centrifuges in parallel are required for appreciable enrichment capacity. There is considerable activity in the development of centrifuge separation facilities, both in this country and in Europe. This scheme utilizes only 10 to 15 percent of the power required by gaseous diffusion, and it appears to be capable of competing quite favorably with the latter method in the overall cost of separative work (see table 13).

Closely related to the centrifuge is the Becker nozzle separation method[16] developed by West Germany. In this device the feed gas is forced at high speed through a curved nozzle. Centrifugal force moves the heavier isotope toward the outer wall where it can be scraped off by a blade. Although nozzle separation is not energy-efficient (even less so than gaseous diffusion), it requires only modest technology aside from the high-precision blade machining.

TABLE 13. Comparative Cost and Performance Factors for
Various Enrichment Methods

	Gaseous Diffusion	Gas Centrifuge	Laser Separation
Separation factor	1.0043	1.5	10
Energy requirement (kilowatt-hour/separative work unit)	2100	210	170
Capital cost (dollars/separative work unit)	388	233	195
Economic size (metric tons)	9000	3000	3000
Process area (acres)	60	20	8
Possible completion date	1985	1982	1986

Source: R. H. Levy, *Science* 192: 866 (1976).

West Germany has recently sold a nozzle enrichment plant as part of a nuclear power package to Brazil. South Africa is also actively developing an enrichment method, the helikon process,[17] which is closely related to the centrifuge and nozzle methods.

Laser Isotope Separation. It has long been known that monochromatic light of the proper wavelength can selectively excite the energy levels of gas atoms or molecules. With the recent development of powerful, tunable lasers, the wavelengths of the light incident on a gas can be tuned to excite selectively the gas atoms or molecules of one isotopic species since there is a slight shift in the energy levels due to the isotopic mass difference.[18] Then the excited species can be separated from the unexcited species by conventional physical or chemical separation methods. For example, in atomic vapor separation the excited atoms can be ionized by a second laser beam and then separated out with electromagnetic fields.

The potential advantages of laser photoexcitation for isotope separation include rather large separation factors (as large as 10), rather modest power requirements, and significantly lower separation costs. The principal disadvantages are the significant technical problems that must be overcome before laser isotope separation can be applied on a commercial scale.

Because of early fears that laser isotope separation technology might be rather simple in comparison with gaseous diffusion or gas centrifuge methods, most research and development on this technique have been highly classified. However, the scaling of the method from its present kilogram capability to the 5,000-ton capacity of a commercial plant will not be straightforward. Laser isotope separation probably will remain a sophisticated technology, and its contribution to nuclear weapons proliferation will be less than that of the more conventional gas centrifuge technology.[19]

Further Comments on Uranium Enrichment. Future needs for enrichment capacity depend sensitively on a number of factors, including the projected estimates of nuclear generating capacity, both domestic and foreign, and the capability of the uranium mining industry to supply feed material. At the present time most of the western world's uranium enrichment requirements are met by the United States's gaseous diffusion plants. Although the capacity of these plants exceeds present demand, by the early 1980s this capacity will be insufficient, and additional enrichment capacity will be needed.

Of course, the need for further enrichment capacity depends on other factors, too. The success of the European gaseous diffusion and gas centrifuge programs will strongly affect foreign demand. Furthermore, since the enrichment output of a plant can be significantly increased by merely feeding in more uranium ore and operating at a higher tails assay, say 0.30 percent rather than 0.20 percent, the capacity of present United States plants can be artificially increased, although at the considerable expense of larger uranium ore feed requirements from an already overburdened mining industry.[20]

The long lead time necessary for plant construction coupled with the projected growth in the nuclear power industry has provided a strong incentive to construct new enrichment capacity. There was a belief for a time that future expansions in capacity should occur in the private sector. However, the enormous capital investments required for such facilities combined with the uncertainty surrounding both future developments, such as gas centrifuge and laser separation technology, and government policy have inhibited the entrance of private industry into uranium enrichment.

Government control of enrichment facilities should not be interpreted as public subsidy of nuclear fuel cycle costs. The federal government charges rather substantial enrichment fees to private industry that amount to a rate of return of about 15 percent per year on the capital value of the enrichment plants. Continued government involvement in uranium enrichment represents a conscious attempt to maintain government control over all technology that could have an impact on nuclear weapons proliferation. The large gas centrifuge addition planned for the Portsmouth plant has been earmarked by the government to supply low-enrichment uranium to foreign nations in an effort to dissuade them from developing independent nuclear fuel reprocessing and plutonium recycling capability.

Fuel Fabrication

The next step in the fuel cycle is the conversion of enriched UF_6 to a solid ceramic or metallic form and then the fabrication of this material into fuel elements. The fuel assemblies in a modern power reactor are extremely complex. Each assembly is not only a source of fission energy, but also a heat exchanger that transfers fission heat to the coolant. It must operate in a severe radiation and thermal environment without failure for a period of several years. These assemblies must be manufactured to fine tolerances to optimize core nuclear and thermal performance. It is understandable why fuel fabrication costs account for almost 20 percent of the total fuel costs.

Reactor Refueling

The fuel assemblies are shipped from the fabrication plant to the nuclear power station for loading into the reactor core. Since they present no radiological hazard at this stage (they contain no fission products), nuclear fuel assemblies can be transported and handled by conventional methods.

Nuclear power units differ dramatically from conventional fossil-fueled plants since they must be shut down and dismantled before refueling can commence. At the designated time of refueling, the reactor is shut down and cooled, and the primary coolant system is depressurized. When the reactor coolant system pressure has been reduced sufficiently, the system is vented, and the coolant level is lowered to a point just below the flange separating the pressure vessel and the vessel head. The control rod drive service lines and other attachments to the head are disconnected, the control rod drives are uncoupled, and the mechanisms holding the head in place are detensioned and removed. The pressure vessel head is then lifted from the vessel. At this time the area above the open vessel is flooded with water to provide a radiation shield. As the upper core support structure is removed, the core is exposed and refueling commences.

The spent fuel assemblies of the core are removed first and transferred to the underwater storage pool where they will be stored for several months before they are shipped off for interim storage or reprocessing. The partially spent fuel assemblies are then transferred to new locations in the core, and the new fuel assemblies are loaded. In a similar manner spent control rods can be replaced and necessary core maintenance can be accomplished at this time. Following refueling and maintenance the reactor is reassembled and then subjected to a series of tests before being started up again.

Spent Fuel Handling, Transportation, and Reprocessing

The fuel assemblies typically remain in the reactor core for a period of three years. During each annual refueling operation roughly a third of the fuel will be replaced. Since the spent fuel removed from the reactor contains an appreciable concentration of fissile material, it is still of considerable value if properly reprocessed. The spent fuel is also highly radioactive, which complicates its handling, shipping, and reprocessing.

Spent Fuel Handling
The spent fuel elements are removed from the core and transferred to water-filled storage pools in the plant and kept for a period of several months while the shorter-lived fission products decay. The underwater storage not only shields against radioactivity, but also removes the considerable decay heat produced in the spent fuel assemblies. Since spent fuel reprocessing has been deferred indefinitely in the United States, high-density storage racks are now being installed in power plants to accommodate discharged spent fuel until interim storage facilities can be constructed.

Spent Fuel Transportation
The spent fuel is next loaded into heavily shielded casks for shipping to reprocessing or storage facilities by either truck or rail. Throughout handling, storage, and shipping the fuel must always be kept in configurations that prevent inadvertent criticality. Of most concern are any accidents in transit that would release radioactivity to the environment.[21] Radiological concerns over spent fuel transport are much different from those characterizing reactor operation. The radioactivity in spent fuel is allowed to decay to a level of only about 1 percent of that of the fuel in the reactor core. Furthermore, since there is no possibility of the fuel element melting in a shipping accident, one worries instead about rupture of the fuel cladding tubes that might release the small fraction of radioactive fission product gases that have diffused out of the pellets and into gaps between the pellet surface and the clad tube.

Spent fuel shipping containers are carefully designed to ensure their integrity in the event of any conceivable accident. Despite the care taken in the design of shipping casks and transportation of spent fuel, there is still great public debate over the shipment of radioactive materials. Unfortunately the distinction is not usually made between the relatively frequent shipments of low-level radioactive materials (over a million such shipments are made every year without incident) and the far less frequent shipments of high-level radioactive materials. For example, fresh nuclear fuel assemblies have low activities and can be touched and handled with bare hands. A typical nuclear plant will discharge roughly two hundred drums of low-level wastes each year that require only modest shielding and can be shipped by ordinary heavy-duty trucks. In sharp contrast a single spent fuel assembly is highly radioactive, requiring that extensive shielding and cooling precautions be taken. Spent fuel shipping containers are designed to withstand normal transport conditions as well as any accident conditions that might arise.[22] Such containers must be able to withstand a crash at 30 mph into a solid wall, envelopment in a gasoline fire for thirty minutes, submersion in water for eight hours, and a puncture test without rupturing. Needless to say, spent fuel shipping casks are massive, complex, and expensive, costing about $2 million each.

Reprocessing of Spent Reactor Fuel
The spent fuel discharged from a nuclear power reactor contains a significant quantity of unused uranium and plutonium. In fact the value of the fissile material remaining in the spent fuel elements discharged each year by a 1,000 megawatt plant is roughly $10 million. The use of this material in new light water reactor fuel elements could stretch uranium ore resources by as much as 40 percent. Hence there is strong incentive to reprocess the spent fuel and extract the unused uranium and plutonium for recycle in power reactors. The reprocessing of spent reactor fuels is also the most logical point at which to extract and concentrate the radioactive fission product wastes into smaller volumes for eventual disposal.

The principal method for commercial recovery of uranium and plutonium from low-enrichment UO_2 light water reactor fuels is a chemical technique known as the Purex process.[23] The spent fuel is received at the reprocessing plant, unloaded from its shipping cask, and stored under water until processing. It is then mechanically disassembled and sheared into short segments. The UO_2 fuel is leached from the cladding and the leached cladding hulls are then rinsed and removed as waste. The dissolver solution, typically nitric acid, is sent through a solvent extraction step that separates more than 99.9 percent of the fission product waste activity from the uranium and plutonium products. The plutonium is extracted as an aqueous plutonium nitrate solution.

The plutonium nitrate solution would next be converted into plutonium oxide powder and returned to a fuel fabrication facility to be blended with uranium oxide to produce new fuel pellets for use in either light water or fast breeder reactors, thereby completing the nuclear fuel cycle.

The technology for industrial-scale spent fuel reprocessing has existed in this country for almost two decades. Extensive experience has been gained in the processing of military reactor fuels, and a small commercial fuel reprocessing facility was operated in West Valley, New York, for a number of years. At the present time a large industrial reprocessing plant stands essentially completed in Barnwell, South Carolina, awaiting only the addition of equipment to convert the plutonium nitrate solution to plutonium oxide and to solidify the fission product waste stream. This plant is unlikely to be put into operation in the near future. The Carter administration decided in 1977 to defer nuclear fuel reprocessing indefinitely because of the implications of this technology for the proliferation of nuclear weapons capability. However, a number of foreign plants are already in operation, including facilities at Marcoule and Cap de la Hague, France, Windscale, England, Tokai-Mura, Japan, and Tarapur, India. A number of other nations have reprocessing plants under construction.

The radioactive effluents from fuel reprocessing plants will probably account for most of the public radiation exposure from the nuclear fuel cycle. Radioactive nuclides such as krypton-85, iodine-129, tritium, and carbon-14 are released in gaseous form from the fuel during the shearing and dissolution operations.[24] If this radioactivity is released, it could eventually contribute an average public exposure of 0.17 mrem per year as compared to exposure from routine power plant emissions of 0.05 mrem per year under the conservative assumption that all electric power generation is nuclear.[25] This is still only 1/500 of the exposure an individual receives from natural sources of radioactivity. Furthermore the technology for removing these gaseous radioactive effluents is presently available, although not yet demonstrated on an industrial level, and therefore the actual releases from future reprocessing facilities will almost certainly be far less than the above estimate.

The recycling of the plutonium produced in power reactors has become an extremely controversial subject. This is due in part to the public perception of the toxicity and sabotage potential of this substance. But it is also due to the implications of nuclear fuel reprocessing for control of nuclear weapons technology.

Since nuclear fuel costs contribute only 10 to 15 percent of the overall electric generating costs, reprocessing and plutonium recycling in light water reactors is projected to have only a modest economic benefit, perhaps as little as 1 percent of the net cost of nuclear-generated electricity.[26] Therefore one might well question why there is a strong incentive to close the fuel cycle by utilizing mixed oxide fuels in light water reactors.

The primary motivation for plutonium recycling is not direct cost savings, but rather the savings in uranium ore feed requirements.[27] By recycling uranium and plutonium in spent fuel elements using even the most conservative recycling scheme (using the same fuel designs for reload cores), one should be able to reduce uranium feed requirements by 30 percent.[28] In the face of our limited uranium resources, such a savings would be quite significant.

Perhaps of even more significance is the need for nuclear fuel reprocessing for future reactor types such as the breeder reactor. The plutonium extracted from light water reactor fuels will be needed to start up the first generation of breeder reactors. Furthermore, fuel reprocessing will be a necessary component of breeder reactor operation. Even though a breeder reactor produces some 30 percent more plutonium than it consumes, its nuclear fuel must still be withdrawn and chemically reprocessed to extract this plutonium so that it can be refabricated into new fuel elements.

A throwaway fuel cycle in which unused uranium and plutonium in spent reactor fuels is discarded as radioactive waste is not an attractive option. In an absence of available reprocessing plants, spent fuel likely will be stored until a decision has been made to proceed with plutonium recycle in light water reactors or to utilize this plutonium in the first generation of fast breeder reactors. Until the mid-1970s it had been assumed that plutonium recycle in light water reactors was desirable, and a Generic Environmental Statement on Mixed-Oxide Fuels (GESMO) was prepared by the Department of Energy to pave the way for this. However, the present moratorium on reprocessing has placed these plans in a hold pattern until further fuel cycle options have been evaluated.

Radioactive Waste Disposal

Probably the most volatile issue concerning the nuclear fuel cycle is the disposal of the high-level radioactive waste produced by nuclear power reactors. A nuclear power reactor will build up an enormous inventory of radioactive fission products during its operation. Although most of this radioactivity will decay away quite rapidly after reactor shutdown and removal of spent fuel elements, a significant fraction of the high-level radioactivity induced in the fuel will remain for many years. To be more precise, let us introduce the concept of *radioactive half-life*, which is the time required for half of the quantity of a given radioactive material to decay away. Since many radioactive materials produced in nuclear reactor fuel have half-lives of decades and longer, this radioactive waste will have to be carefully isolated from the environment for a significant period of time.

In fact, since the radioactive waste produced by power reactors will contain minute quantities of transuranium isotopes such as plutonium with radioactive half-lives as long as tens of thousands of years, it is commonly claimed that such wastes will have to be isolated and guarded in perpetuity. This claim is frequently stated with an air of indignation at the immorality of leaving a legacy of radioactive waste as a potential hazard for future generations. Perhaps the fundamental concern in the generation and disposal of radioactive waste is that in pursuit of our own energy and security we may be responsible for inflicting harm on future generations by failing to isolate this radiologically hazardous material adequately.[29]

To place some of these concerns in perspective, we must first recognize that there will be a radioactive waste disposal problem whether we rely on nuclear power generation in the future or not. Over three decades of processing the fuel from military reactors such as those used to drive nuclear submarines or to produce the material for nuclear weapons has already generated a significant volume of high-level wastes. These are presently stored in either liquid or solid form on government reservations in Washington and South Carolina. It is estimated that the present volume of military wastes is perhaps ten times the amount that will be generated by the commercial nuclear power industry by the turn of the century.[30]

The *physical amount* of radioactive waste produced by nuclear power reactors is actually quite small. For example, the volume of high-level waste produced during a year's operation of a large power reactor is only about 2 cubic meters.[31] All radioactive waste that will be generated by the entire United States nuclear power industry until the year 2000 would fit into a cube about 80 meters on a side, and of that, the high-level wastes would occupy a cube only about 15 meters on a side. When compared with the enormous volume of wastes produced by other types of electric generating plants (it takes a train of thirty-three carloads a day just to remove the ashes from a modern coal plant), it is apparent that the volume of radioactive waste is actually quite small.

The principal concern involves the *toxicity* of radioactive wastes. They must be disposed of in a manner that ensures that they are isolated from the human environment long enough to decay to harmless levels. Although many radioactive materials are produced in nuclear reactors, most activity in radioactive wastes is due to fission products such as strontium-90 and cesium-137, which have radioactive half-lives of about thirty years and will decay to harmless levels in several hundred years (or ten half-lives). However, radioactive wastes will also contain trace amounts of heavy radioactive elements such as the actinides plutonium, neptunium, and americium, which have half-lives of thousands of years. Therefore radioactive wastes will exhibit a residual radioactivity for a much longer period.

The magnitude of this very slowly decaying transuranium component is usually rather small, however. For example, the processed high-level radioactive wastes from a nuclear power plant will decay to essentially the same radiotoxicity level as natural uranium ore after several hundred years.[32] Therefore the toxicity of such wastes drops dramatically over this period. One would have to eat several hundred grams of such wastes to do himself bodily harm after this time.[33] Contrast this with toxic substances such as arsenic or lead, which we reject in an uncontrolled fashion into our environment and which will *never* decay to a benign level. Coal also contains trace amounts of radioactivity with average uranium and thorium concentrations of several parts per million.[34] The concentrations measured in the flyash of power plants ranges as high as 40 parts per million. After five hundred years, the toxicity of high-level nuclear waste is less than that of coal ash containing 24 parts per million uranium.

The present plan[35] for treating high-level radioactive wastes involves storing them as a liquid waste solution after fuel reprocessing for a period of up to five years in large underground storage tanks. This allows the bulk of the fission product activity and heat generation to decay to substantially lower levels and facilitates handling safety and chemical processing during further treatment of the waste. The storage tanks have double-lined stainless steel walls and are monitored continually to prevent any leakage to the environment. These tanks are much different from the old World War II tanks containing military wastes at Hanford that did experience some leakage problems.

Federal regulations require that liquid waste solutions be converted to stable solids such as glass or cement and encapsulated in

water-impervious inert containers within five years of their generation and shipped to a federal waste repository within another five
years. The most attractive plan for handling the liquid waste from
fuel reprocessing facilities is to calcine the waste by high-
temperature drying and conversion to an oxide form. The resulting
calcine powder is then mixed with powdered glass, and the mixture
is melted, cast into a glassy form, and sealed in stainless steel
canisters, about 0.3 meters in diameter and 3 meters in length. These
canisters would then be shipped to a federal nuclear waste repository
for first temporary or retrievable storage. They would eventually be
transferred to a terminal or permanent storage facility (see fig. 18).

The most promising scheme for permanent disposal of high-
level solidified wastes is geologic emplacement.[36] The waste canisters would be buried at depths of 600 to 1,000 meters beneath the
earth in rock formations that exhibit exceptional geologic stability.
Although most attention has been directed at salt formations that
have remained geologically undisturbed for millions of years, some
consideration has also been given to granite, shale, and basalt forma-

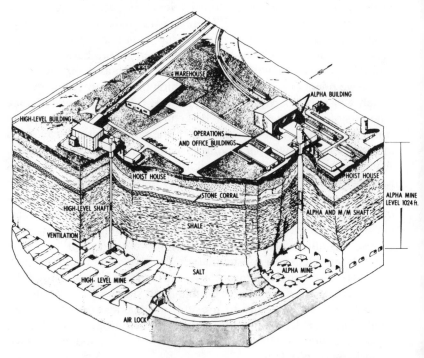

Fig. 18. A radioactive waste repository. (*Courtesy of the United States Department of Energy.*)

tions. A relatively small amount of land would be required for such a waste repository. For example, a typical power plant would produce some ten canisters of high-level waste each year. If these canisters are stored in rows, 10 meters apart, then only about 10 square miles would be needed to accommodate all of the radioactive wastes produced over the next one thousand years.

Geologic disposal raises the question of the hydrogeology of the site.[37] If groundwater came into contact with the waste, it could eventually be leached into solution, move through aquifers with the water, reach the surface, and contaminate the food chain. However, the time scales for geologic change are extremely long. Even though there has been some concern regarding the presence of water in salt formations[38] (a concern that led to the abandonment of an early waste disposal demonstration site at Lyons, Kansas), there seems to be general agreement that the hazards posed by geologic emplacement are extremely small even when human intervention such as by inadvertent drilling is taken into account.

Various alternative methods of permanent waste disposal have been suggested.[39] Perhaps the most attractive alternatives involve rock melting and superdeep drilling and emplacement. Alternative geologic sites, such as beneath the ocean floors or polar ice caps, have been suggested. More exotic suggestions include using space vehicles to jettison wastes from the solar system (highly expensive and hazardous) or transmuting the long-lived radioactive wastes (the actinides), into stable or short-lived waste products by inserting them into high neutron flux reactors.[40] However, the consensus of informed scientific groups is that effective long-term isolation for spent fuel, high-level, or transuranic waste can be gained by geologic emplacement.[41]

Several other comments should be made concerning radioactive waste disposal. Almost regardless of the method chosen for permanent disposal, the costs of handling and disposing of the waste are expected to be quite small relative to the costs of generating electric power. Electric utilities are charging 0.5 mills per kilowatt-hour on their books to pay for eventual radioactive waste disposal.[42] They are also including a charge of 0.2 mills per kilowatt-hour to cover eventual costs for decommissioning the power plant. That is, after its useful operating lifetime of some thirty to forty years, a nuclear power plant will contain radioactive components that must be dismantled and disposed of in some suitable fashion. The decommissioning of nuclear power plants is projected to cost some $20 to $40 million—a modest expense compared with the capital cost of the plant and the revenue generated over the plant's lifetime.

In summary, there seems to be little doubt that present technology can provide for the safe disposal of radioactive waste. However, it is one thing for a scientist to recognize the existence of this technology, and quite another thing to demonstrate it convincingly to the public. Although radioactive waste repositories are under development in Germany, France, and the Soviet Union, the United States program has ground to a halt. Government inaction on the development, licensing, and operation of demonstration facilities has given the impression that there are no solutions available for disposal of radioactive waste. This impression has been compounded by the decision of the Carter administration to defer indefinitely spent fuel reprocessing, since the "stowaway" fuel cycle demands interim storage of spent fuel rather than permanent disposal of radioactive wastes.

A number of states have blocked further nuclear power development until permanent radioactive waste disposal technology has been demonstrated to their satisfaction, an action they have taken in spite of assurances from scientific groups that there are no technical barriers to the development of waste repositories on a timely basis.[43] The states are apparently ignoring the fact that significant quantities of radioactive wastes will not be discharged from nuclear power reactors or fuel reprocessing plants for several decades. In this period the overwhelming contribution to radioactive wastes will be those already accumulated by the military weapons program.

Plutonium

The nuclear fuel cycle involves appreciable quantities of the artificial element plutonium. Occasionally plutonium is referred to as "the most toxic substance known to man."[44] If this statement were true and if the probability of plutonium being released into the environment and finding its way to man were significant, then this in itself would be a major argument against the massive implementation of nuclear power.

Plutonium is indeed highly toxic, although certainly not the most toxic substance in our society.[45] The high toxicity of plutonium is not due to its chemical properties as is the case with other highly toxic materials in our environment such as arsenic or cadmium, but rather its radioactivity. Furthermore, since this radioactivity is in the form of alpha radiation that cannot penetrate the skin, plutonium must be ingested or inhaled to be hazardous. Fortunately plutonium is not readily absorbed by the body. The absorption of plutonium through the skin is small, although absorption through wounds or abrasions can be a problem. Ordinarily any plutonium that enters the body through the digestive tract will not be retained. The amount of plutonium one would have to ingest to have a 50 percent chance of dying is some 3 grams, only a tenth of the ingestion toxicity of materials such as arsenic or lead. Inhalation of plutonium into the respiratory tract is the major mechanism by which plutonium enters the body. For this reason most concern is directed at the tendency of plutonium or other actinides to damage lung tissue.

There have never been any known fatalities attributable to plutonium poisoning, although large numbers of plutonium workers have been monitored for several decades to assess the effects of plutonium on health. Most of the data on the radiotoxicity of plutonium have been obtained from experiments performed on animals. Nevertheless we probably know more about the toxicity of plutonium by the means of such studies than we know about any other element in the periodic table.[46] These studies have demonstrated that plutonium inhalation can increase the risk of lung cancer. Therefore stringent radiation protection standards have been adopted that limit the permissible level of plutonium (the body burden) present in members of the general public or in employees handling the material.

Recently some concern has been voiced that radiation standards for plutonium are too high. One particular theory that received great public attention asserted that plutonium particles of a certain size are one hundred thousand times more hazardous than the same amount of plutonium distributed uniformly throughout tissue.[47] If this so-called hot particle theory of plutonium toxicity were true, then the permissible lung burden limits would have to be lowered significantly. But the hot particle theory was rejected by essentially every scientific group that studied plutonium toxicity, including the National Council on Radiation Protection and the National Academy of Sciences.[48] Experiments seem to indicate that uniform distribution of radiotoxic material is far more likely to induce tumor formation than concentrations of radioactive material in the form of particles.

It is sometimes suggested that the dispersal of plutonium in the environment by a terrorist, assuming that one could obtain a quantity of this substance, would pose a significant threat to society. However, this would not be a particularly effective form of terrorism.[49] For example, the dispersal of one pound of reactor grade plutonium in a city would cause roughly twenty-five cancers eventually. Of course, these cancers would probably not appear for many years. The dispersal of plutonium in the ventilation system of a building could be somewhat more serious, but again there would be no immediate consequences from this act, only an increase in the probability that exposed individuals would eventually develop lung tumors. Many other toxic substances could be dispersed with far more dramatic and immediate consequences.

To place the concern over plutonium toxicity in perspective, we must recognize that all of us have some plutonium in our bodies, and in our lungs in particular. During the days of atmospheric nuclear weapons testing, over 6 tons of plutonium were dispersed into the atmosphere. Moreover, we are continually exposed to other radioisotopes of comparable toxicity that occur in nature. For example, all of us are exposed to radon, which has a radioactivity per unit mass about five times that of plutonium. Furthermore, those who smoke inhale significant quantities of the radioactive isotope polonium-210 from tobacco. There has been some speculation that the presence of this radionuclide in cigarette smoke may be one cause of lung cancer in smokers. Finally we must keep in mind that all of us are exposed to a natural background radiation level of some 100 mrem per year. The radiation exposure of the general public from plutonium is negligible by comparison.

Certainly the toxicity of plutonium requires stringent methods of control, particularly as the inventories of this material build up in the nuclear reactor fuel cycle. However, some thirty years of experience have indicated that plutonium can be safely handled and isolated from our environment.

Nuclear Power, Sabotage, and the Bomb

Engage any of your friends in a conversation about nuclear power and you are bound eventually to encounter statements such as
   "Nuclear power plants are easy to sabotage."
   "Today the knowledge and equipment required to make atomic bombs are easy to obtain. Any competent scientist or engineer (even a Princeton undergraduate) can build one in his basement."
   "To protect against nuclear sabotage or the theft of nuclear materials, we will be forced into a police state."
The popular media have implanted in the public mind the image of nuclear power plants churning out tons of weapons grade plutonium that can be assembled rather easily into thousands of atomic bombs. Such an impression is most unfortunate and incorrect.

Sabotage of Nuclear Power Plants
We must ask ourselves just what a group of terrorists could do to a nuclear power plant.[50] Of course, they could disrupt its electric

power output and cause substantial damage to the plant, although a nuclear plant is far less vulnerable to terrorist attack in this respect than a fossil-fueled plant or a hydroelectric dam, for example. But could they endanger the public?

A nuclear reactor cannot be made to explode like a bomb. The principal concern would be whether sabotage could cause a release of radioactive fission products to the environment. To accomplish this, saboteurs first would have to gain access to the plant by breaking through security barriers and overcoming a security force. Then they would have to disable all engineered safeguards designed to prevent fission produce release and initiate a loss of coolant accident. These latter tasks would require significant technical expertise. Even if these tasks could be accomplished, the probability that substantial public damage or injury would occur is still quite small. Here we recall that the WASH-1400 Reactor Safety Study indicated that the most probable consequences of a loss of coolant accident were quite minimal.[51]

Of course the terrorist might dream up more indirect schemes such as bombing the plant or crashing a plane into it. But the reactor containment is a massive structure built to withstand major impacts such as those due to airplane crashes or tornadoes, and the explosive force necessary to penetrate this containment would no doubt cause far more destruction if dropped directly on a population center than any fission product release that might ensue from bombardment.

It soon becomes evident that our society contains far more vulnerable targets for terrorist attack than nuclear power plants. For example, it is rather easy to plant a bomb in a large building or public place. Public water and food supplies are quite vulnerable. (One city was threatened with having fuel oil dumped in its water supply.) Furthermore there are massive dams, football stadiums, airplanes, and so on, the sabotage of which would undoubtedly result in far greater consequences than the sabotage of a nuclear power plant.

Our society has faced the ever-present threat of terrorism without downgrading technology, although occasionally additional security precautions have been taken, as in airport inspections. To refuse to implement a new technology simply because of the threat of terrorism (and, in the case of nuclear power, a rather small threat at that) is tantamount to surrendering our sovereignty and freedom of action to those who would use such means to achieve their goals.

The Amateur Atomic Bomb Builder

The myth that any competent scientist or engineer could build an

atomic bomb in his basement, provided he could obtain the necessary materials such as plutonium, after only a few visits to his local
library and hardware store has pervaded much of the debate over
nuclear power in recent years.[52] It is important that we examine this
topic in some detail.

First, what kind of bomb-stuff is needed for an atomic bomb?
The ideal materials are either pure uranium-235 or plutonium-239
metal, but the oxide form of either isotope would suffice, as would a
less than 100 percent concentration. To give some idea of the
amount of material required, we have listed in table 14 the amount of
each type of material necessary to achieve a critical mass[53] (but not
necessarily to fabricate a weapon). We have not bothered to include
uranium with enrichments below 10 percent since this low concentration of U-235 could not be made into an explosive device, regardless of how clever one is.

The next question is, Where does the amateur bomb builder get
this material? The obvious answer is the nuclear fuel cycle—unless
he has the nerve to break into a military weapons depot, which is
entirely unrelated to nuclear power development. We have sketched
a simple diagram of the nuclear fuel cycle in figure 19 to indicate
those points at which "strategic nuclear materials," an official
sounding name for bomb-stuff, could be diverted. As we have noted,
the 3 percent enriched uranium utilized by light water reactors is too
dilute for explosive devices, so we can rule out U-235 as an available
material. The other alternative is the plutonium produced in the
reactor. Although an appreciable quantity of plutonium is produced
during the operation of a power reactor, we must remember that this
plutonium comprises only 1 percent of the irradiated fuel and therefore must be separated out before it will be of any use. Hence the
spent fuel discharged from power reactors is useless as bomb material. Only the plutonium separated out during fuel reprocessing is

TABLE 14. Critical Mass Required for Various Types of Fissile
Material (Spherical Geometry Surrounded by a Natural
Uranium Reflector)

Type of Material	Critical Mass (in kilograms)
Pure Pu-239	4
Pure U-235	15
Reactor grade Pu	8
20 percent enriched uranium	250
10 percent enriched uranium	1300

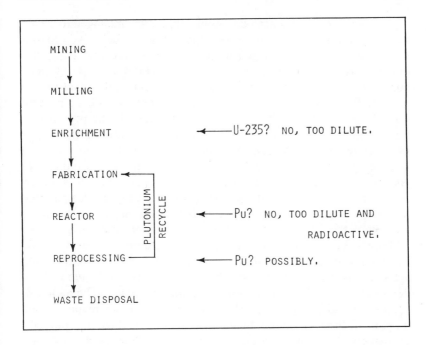

Fig. 19. Possible sources of strategic nuclear materials in the nuclear fuel cycle.

suitable. The only vulnerable stage of the nuclear fuel cycle is the spent fuel reprocessing operation.

Of course, there are many ways to thwart the diversion of plutonium at this stage. For example, we could locate fuel fabrication and reprocessing plants on the same site so that the extracted plutonium is immediately mixed back in with slightly enriched uranium to make fresh fuel elements. We could modify the reprocessing operation so that plutonium and uranium are never separated (so-called *coprocessing*). Or we could denature the plutonium extracted from spent fuel by mixing it with strong neutron absorbers or emitters or intensely radioactive materials that would make it more difficult to use in weapons fabrication. Perhaps the simplest approach is to harden both reprocessing facilities and transportation vehicles by adding suitable security systems and accounting procedures. For some time now United States Army Special Forces teams have been used to test the security of military plutonium storage facilities and transportation vehicles. The safeguards to prevent diversion of strategic nuclear materials from the military nuclear weapons program are already quite extensive, and these could be readily adapted to civilian reprocessing activities.

Suppose that our basement bomb builder has managed to obtain some plutonium from reprocessed spent reactor fuel. Even then it would be difficult for him to use this in an explosive device because of the isotopic composition of reactor grade plutonium.[54] The typical composition of plutonium discharged from light water reactors is 2 percent Pu-238, 58 percent Pu-239, 25 percent Pu-240, 11 percent Pu-241, and 4 percent Pu-242. The first problem is the presence in reactor grade plutonium of an appreciable amount of the highly radioactive isotope Pu-241. Handling this material would present a significant radiological hazard. Furthermore both Pu-240 and Pu-241 give rise to large, spontaneous neutron emission rates. It is important to minimize stray neutron production in a weapons design since this might trigger the chain reaction before the pieces of the bomb are completely assembled into their most supercritical configuration. Although this feature of reactor grade plutonium can be overcome by careful design, it is highly unlikely that the amateur would achieve this degree of sophistication. In fact, the explosive yield of devices fabricated out of reactor grade plutonium would be quite low (several hundred tons of TNT equivalent). Homemade atomic bombs, if they could be built, would be blockbusters, not city destroyers.

Parenthetically weapons that have been fabricated out of plutonium, such as the device exploded by India in 1974, utilized plutonium produced in low-power reactors that could be refueled in such a way as to minimize the buildup of Pu-240 and Pu-241. For example, in heavy water reactors one can continuously refuel, thereby achieving only short exposures of U-238 and producing primarily the lower mass isotope of plutonium, Pu-239. In light water reactors refueling is conducted at yearly intervals; hence a significant buildup of Pu-240 and Pu-241 cannot be avoided without tampering with the refueling schedule.

The amateur bomb builder will have to face still other hurdles. The fabrication of a nuclear weapon from even optimum materials such as Pu-239 is difficult and requires sophisticated electronics and circuitry, explosives, machining, fabrication, and handling. Furthermore handling plutonium (particularly reactor grade plutonium) is a dangerous game, particularly if you do not know what you are doing. The radioactive heat generated by plutonium is enough to melt most plastic explosives used in nuclear devices. Furthermore the bomb design is extremely sensitive to the isotopic composition of the plutonium, and this composition is usually not known to the accuracy required for weapons design for spent fuel plutonium without extremely sophisticated measurements. Finally, contrary to popular belief, much of the information involved in fission weapons design is not available in the open literature, and most of the information that is unclassified is still rather hard to come by.

The highly publicized "workable designs" of atomic bombs produced by amateurs such as college students and newspaper reporters must be classified as schematics, not as actual designs. A design must contain complete technical specifications of the device sufficient for its fabrication. Furthermore the manufacture, as opposed to the design, of a nuclear bomb is a complex operation demanding considerable effort and continued success through a number of difficult steps. Although a dedicated individual could conceivably design a workable device, building it would be an entirely different matter. Only a group with knowledge and experience in explosives, physics, and metallurgy, and the requisite financial resources and nuclear materials could build a crude nuclear explosive.[55]

Every nation that has developed nuclear weapons has devoted the efforts of thousands of people including scientists, engineers, and technicians to the bomb development. We must therefore conclude that the connection between nuclear power and plutonium for bombs is very tenuous, because of the heavy security surrounding the limited sites where reactor grade plutonium is present in satisfactory form, because of the unsuitability of this form of plutonium for explosive devices, and because of the greatly underestimated difficulty of fabricating such devices. Because of the present moratorium on reprocessing spent power reactor fuel in the United States, there is *no* strategic nuclear material in the stowaway fuel cycle suitable for nuclear weapons. Therefore the fears of subnational plutonium diversion and basement bomb building appear to have been greatly exaggerated.

Closing the Nuclear Fuel Cycle

It is apparent that the nuclear fuel cycle in this country will not be closed for many years. Although the technology involved in spent fuel reprocessing, plutonium recycle, and radioactive waste disposal has been clearly demonstrated for some time, the commercial implementation of these activities faces a number of major social and political barriers. The debate over plutonium recycle and radioactive waste disposal has become so emotional and wandered so far from technical issues that the federal government has drifted into a position of actually blocking the further development of the commercial nuclear fuel industry. The absence of suitable licensing and regulatory procedures is particularly serious.

The identification of nuclear fuel reprocessing as an international issue and the subsequent decision to defer indefinitely plans to implement this technology have contributed to the uncertainty surrounding the nuclear fuel cycle. It seems that the relationship between nuclear power development and the international implications of spreading nuclear technology now plays the dominant role in determining the stance of the United States toward nuclear power. Therefore it is essential that we examine the international aspects of nuclear power in some detail.

5

International Aspects of
Nuclear Power

One customarily thinks of the United States as the world leader in nuclear power development. However, early in 1976 the rest of the world surpassed the United States in nuclear generating capacity. Today, even as the United States nuclear power program has been slowed to a crawl by public controversy, government indecision, and adverse economic conditions, the worldwide commitment to nuclear power development continues to grow at a rapid rate. During the next decade United States nuclear activity, whether measured in generating capacity, uranium production, or enrichment or reprocessing capacity, will shrink to less than 25 percent of the worldwide total.

Rapid escalation in the cost of fossil fuels coupled with the problems resulting from a growing reliance on petroleum imports (dramatized by the oil embargo of 1973–74) has generated intense international pressures to develop alternative sources of energy. Since most nations have limited domestic energy resources, the incentive to implement nuclear power technology rapidly has been strong. Without nuclear alternatives, most industrialized nations face not only declining living standards, but also a balance of payments deficit caused by massive foreign oil imports.

**Present Status and Trends in International
Nuclear Power Development**

At the present time some fifty-two countries have made firm commitments to nuclear power development.[1] (See table 15.) Twenty of these nations currently depend on nuclear power generation to supply a significant fraction of their electric energy needs. A number of European nations, including Belgium, Sweden, Switzerland, the

TABLE 15. Survey of Nuclear Power Plants

Country	Operating	Under Construction	Ordered[a]	Planned[a]	Total[a]
Argentina	1	1	3		5
Austria		1		1	2
Belgium	3	2	2		7
Brazil		3		6	9
Bulgaria	2	1	1	4	8
Canada	7	10	4	5	26
China (Taiwan)		4	2		6
Czechoslovakia	1	4		16	21
Denmark				6	6
Egypt				5	5
Finland		4			4
France	10	17	12	8	47
German Democratic Republic	3		2		5
German Federal Republic	7	12	8	4	31
Hong Kong				1	1
Hungary		1	1		2
India	3	5			8
Indonesia				3	3
Iran			4	1	5
Ireland				1	1
Israel				1	1
Italy	3	2	4	16	25
Japan	10	14		5	29
Korea (South)		1	1	8	10

Luxembourg			1		1
Mexico		2		7	9
Netherlands	2			3	5
Pakistan	1			1	2
Philippines			2	8	10
Poland				2	2
Portugal				4	4
Romania			1	2	3
South Africa			2		2
Spain	3	7	7	21	38
Sweden	5	6		3	14
Switzerland	3	1	3	2	9
Thailand				3	3
Turkey				1	1
USSR	19	8		10	37
United Kingdom	29	10		7	46
Yugoslavia		1		1	2
Cuba[b]				8	8
Kuwait[b]				2	2
Libya[b]				2	2
New Caledonia[b]				2	2
45 countries	112	117	60[a]	180[a]	469[a]

Source: Atomic Industrial Forum, 1977.

a. Forecasts of future nuclear capacity tend to be highly uncertain, since they depend quite sensitively on both economic and political factors.

b. No details are available on the implementation of the planned nuclear power program for these countries.

United Kingdom, and France, already derive over 15 percent of their generating capacity from this source.

This trend toward nuclear power is likely to intensify in future years as fluid fossil fuel reserves are rapidly depleted. The nuclear commitment of industrialized nations will be imitated by developing nations, which will require a larger and larger fraction of world energy generation in the future.[2] Hence, even if we assume that industrialized nations can implement drastic energy conservation measures to limit energy consumption, we find that worldwide energy demand will continue to grow at a significant rate. Most nations of the world do not possess significant fossil fuel resources and cannot rely in the short term on unproven alternatives such as solar energy or fusion power. It is understandable that nuclear power is commonly perceived as playing a significant role in their future.

The United States has played the leading role in the international development of atomic energy since the Atoms for Peace program initiated by the Eisenhower administration in 1954. This program paved the way for the international transfer of nuclear power technology, subject to controls administered by the International Atomic Energy Agency. The United States continued to dominate the international nuclear power industry through technical assistance programs and the sales of nuclear power equipment until the early 1970s. Stimulated by this activity, many nations made strong commitments to nuclear energy. Table 16 gives a selective chronology of international nuclear power development.[3]

Even as the international development of nuclear power matured, the role of the United States as an international supplier of nuclear power technology began to fade. The dramatic changes in the domestic and foreign policies of the federal government regarding nuclear power have led to a sharp decline in its influence in the international market. The inconsistency and confusion in the United States's energy policy and leadership, particularly in nuclear power development, have stimulated a number of nations to build and operate nuclear power plants of their own and develop the entire nuclear fuel cycle necessary to support these plants. In addition to the United States, Great Britain, the USSR, France, West Germany, and Canada now have full capabilities for nuclear power development. All of these nations are formidable competitors with the United States in the international market for nuclear power technology.

Seven nations have entered the international market as suppliers of nuclear reactors, equipment, and fuel. Four nations

TABLE 16. Highlights of International Nuclear Power Development

1940s	Viewed as inexhaustible resource with breeders
1954	U.S. Atoms for Peace program—conditions of nonproliferation treaty in bilateral agreements Purex reprocessing technology exported by U.S.
1956	Suez Crisis—breeder given emphasis in U.K. and France as substitute for oil-fired plants U.K. and France have reprocessing capability
Late 1950s	First commercial plants order from U.S. by West Germany and Italy
1958	First financing of nuclear power by World Bank
1960-66	First commercial plants ordered by France, India, Netherlands, Japan, Spain, Switzerland
1967	Arab-Israeli War—triggered orders and breeder demonstration programs
1968-73	Steady increase in commercial orders
1974	Arab oil embargo causes flurry of orders (many moved up from later planned dates)
1973-present	General increase in orders, development of breeder Orders not going to U.S. suppliers due to lack of assured fuel supply Negotiations by London Nuclear Suppliers Group to place safeguards on nuclear technology transfer

Source: B. A. Hutchins, "Nuclear Energy Programs in Other Nations" (Paper presented at AUA-ANL Conference on International Aspects of Nuclear Power, Argonne National Laboratory, Argonne, Ill., May 16, 1978).

besides the United States have commercial scale enrichment facilities in operation (the United Kingdom, France, the USSR, and China), and an additional six nations are constructing or planning commercial enrichment plants. Commercial fuel reprocessing facilities exist in five nations outside the United States (the United Kingdom, France, the USSR, Belgium, and Japan), and a number of other nations have pilot reprocessing facilities in operation. See table 17.

Obviously the United States no longer has a monopoly on commercial nuclear power technology. Other industrialized nations have developed the capability to compete in the international nuclear power market. This has given rise to several controversial actions such as the proposed sale of reprocessing technology to South Korea, Pakistan, and Iran by France[4] and the sale of a complete nuclear power industry, including power plants, enrichment, and reprocessing facilities, to Brazil by West Germany.[5]

TABLE 17. Commercial Nuclear Power Capability of Industrialized
           Nations

Nation	Nuclear Equipment Manufacturing	Uranium Enrichment	Spent Fuel Reprocessing
Belgium	—	Under const.	X
Canada	X	—	—
China	X	X	X
France	X	X	X
India	X	—	X
Japan	X	—	X
South Africa	—	X	—
Sweden	X	—	—
United Kingdom	X	X	X
United States	X	X	X
USSR	X	X	X
West Germany	X	Under const.	Under const.

The impetus for future nuclear power development has shifted from the United States to Europe as the United Kingdom, France, the USSR, and West Germany have taken significant steps toward the development of fast breeder reactors. These nations have also made significant progress toward closing the nuclear fuel cycle by building and operating commercial scale fuel reprocessing facilities and developing radioactive waste disposal technology and repositories. The degree of international cooperation and the advanced nature of the nuclear power development programs exhibited by these nations suggest strongly that Europe will dominate the nuclear power market for some time to come.

A Brief Survey of Nuclear Power Development
in the Major Industrialized Nations
The United Kingdom has been heavily committed to nuclear power development since the Calder Hall nuclear power station went into operation in 1956.[6] Such natural uranium, CO_2-cooled Magnox reactors have supplied roughly 10 percent of Britain's electric generating capacity for the past twenty years. The high capital cost of this type of nuclear plant has motivated serious consideration of alternative types such as the advanced gas-cooled reactor (AGR), the steam-generating heavy water reactor (SGHWR), and even the imported light water reactor. However, the recent discovery and exploitation of North Sea oil reserves has reduced the pressure on Britain to expand its nuclear generating capacity rapidly, and therefore it will probably not commit to an alternative reactor type for several years. Nevertheless the British have constructed and suc-

cessfully operated a demonstration fast breeder reactor, the Prototype Fast Reactor.

Great Britain is also moving rapidly toward closing the nuclear fuel cycle. Although it already reprocesses spent fuel from Magnox reactors, it is presently increasing the capacity of its Windscale reprocessing plant to handle both future domestic spent fuel requirements and light water reactor fuels of other countries, most notably Japan. It has also collaborated with several European nations in the construction of a gas centrifuge plant at Capenhurst.[7]

France also was an early pioneer in nuclear power development. The first patents on fission chain reactors were obtained by French scientists. Early French activity was based on gas-cooled natural uranium reactors. However, with the development of an independent uranium enrichment capability (originally stimulated by the French nuclear weapons program), France shifted toward pressurized water reactors. Early pressurized water reactor plants were built by the French firm Framatome, operating under a Westinghouse license. France today stands as one of the world's leading suppliers of nuclear power equipment.

The French government has made a significant commitment to nuclear power development. By the turn of the century, France plans to derive more than 90 percent of its electricity from nuclear power. In fact, it intends to build no more fossil-fueled generating plants. It has moved aggressively toward the development of an independent uranium enrichment and spent fuel reprocessing capability. Furthermore, encouraged by the success of the Phenix demonstration fast breeder reactor at Marcoule, France has under construction a commercial-size (1,200 megawatts) breeder reactor, the Superphenix,[8] at Creys-Malville (see chapter 6). It plans to begin commercial sales of this reactor type in the early 1980s. There is little doubt that France has become the world leader in nuclear power development due to both the breadth and sophistication of its nuclear power industry as well as the degree of its commitment to this energy source.

West Germany has made a significant commitment to nuclear power. It was recognized early that Germany's recovery after World War II would depend heavily on adequate supplies of energy and that, because of its own limited fossil fuel reserves, nuclear power would play an important role in meeting Germany's energy requirements. The German nuclear power program has moved rapidly along a number of fronts. Today West Germany is regarded as a world leader in light water reactor technology. Germany also has a strong program in gas-cooled reactor development and fast breeder reactor technology. It has become a dominant force in the international market, a fact that was reinforced by its $5 billion sale of a complete nuclear power industry to Brazil.

The Soviet Union has moved more deliberately toward nuclear power deployment. It now plans to have a nuclear capacity of about 20 gigawatts on-line by 1980, roughly one third of the present United States generating capacity. Although the USSR has significant fossil fuel reserves (principally coal), these resources are located far from major population centers. Therefore nuclear power is expected to play a significant role in the future Soviet energy program. The Soviet Union is a major exporter of nuclear power equipment and has had a major fast breeder reactor development program underway for a number of years.

The Canadian nuclear power program is based primarily on heavy water, natural uranium fueled (CANDU) reactors.[9] Canada has developed an independent capability to manufacture, fuel, and export this reactor type. In fact, the first international sales of nuclear power reactors were made by Canada to India. The Canadian government is moving aggressively toward international marketing of the CANDU reactor while it rapidly expands its own domestic nuclear generating capacity. It has even been proposed that Canada build a chain of nuclear power stations along the United States border to produce electricity for sale to the northeastern United States.

Other industrialized nations have implemented ambitious nuclear power programs: Japan, Sweden, Belgium, Italy, Spain, and the Warsaw Pact nations. In fact, there are few industralized nations in the world that have not made a significant commitment to this energy source.

The Role of Nuclear Power in Meeting the
Energy Needs of Developing Nations
Many developing nations also have moved rapidly to embrace nuclear power. The emergence of an intensely competitive international market in nuclear equipment, coupled with the staggering sums of money generated by rapidly rising petroleum prices (petrodollars), has permitted a number of nations (particularly OPEC nations) to consider the acquisition of nuclear power technology. For example, Iran publicized its intentions to have some thirty nuclear plants on-line by the turn of the century, which would rank it as the fourth leading nation in nuclear power capacity. Subsequent political upheaval in early 1979 has forced Iran to curtail these plans drastically. Brazil has a contract with West Germany to acquire

from two to eight large nuclear power stations plus the technology for uranium enrichment and spent fuel reprocessing.

India has also mounted an aggressive nuclear power program. After acquiring boiling water reactors from the United States and heavy water reactors from Canada, India has continued on to develop the capability to manufacture and fuel CANDU-type reactors. Six such plants ranging from 200 to 240 megawatts are either in operation or under construction. India may even become a major supplier of nuclear power equipment to third world nations. India has also developed the necessary nuclear fuel cycle technology, including spent fuel reprocessing, to support its growing industry.

As the energy needs of such developing nations continue to grow, it is reasonable to expect that many will move rapidly to acquire nuclear power technology.

Projections of Future International
Nuclear Power Development

There is little doubt that the significant role that the United States has played historically in the international development of nuclear power will diminish in the future. This will occur because of the rapid expansion of worldwide nuclear power capacity and because of a number of internal political decisions that have significantly inhibited the growth of nuclear power deployment in this country.

One major factor in recent decisions to throttle back on nuclear power development in this country, particularly in the areas of spent fuel reprocessing and breeder reactor development, has been the concern about international nuclear weapons proliferation.[10] Although this unilateral slowdown of the United States nuclear power program was meant to persuade other nations to limit the further spread of nuclear technology, it has had the opposite effect in some instances.[11] As those nations that have already made a significant commitment to nuclear power development perceive a weakening of the United States position as a potential supplier of nuclear fuels and services, they feel even more compelled to develop independent capabilities in these areas. Furthermore there is certainly a tendency among nations who are competitors of the United States in the international commercial nuclear power market to view with suspicion any proposal by this country that they should limit their own nuclear power industry. Finally they see the United States continue to place enormous pressures on world oil reserves by massive foreign oil imports, even as it slows the development of nuclear technology that could reduce these imports to some degree.

The public controversy in the United States over nuclear power has created strong pressures on public officials and regulatory agencies to slow the development of this energy source. The same controversy has spread to a number of other nations, including France, West Germany, Sweden, and Austria.[12] The opposition to nuclear power development has been expressed through political channels such as referenda (Austria) and parliamentary elections (Sweden). It has also occasionally taken on a violent tinge as radical groups have seized on demonstrations against nuclear power as a mechanism for disrupting government policy and forcing change. Already there is increasing evidence that such political activities have significantly hindered nuclear power development in Europe. Of course, the nuclear power programs of authoritarian countries in which such protests are not allowed have proceeded without interruption.

Even if nuclear power generation represents a viable and attractive energy option for most industrialized nations of the world, there is still considerable uncertainty about nuclear power as a desirable route for developing nations. Nuclear power generation requires a significant investment in both economic and manpower resources. This may well be beyond the capability of many nations that have recently considered a nuclear future. The transfer of such high technology to an underdeveloped country characterized by primitive industrial capability is fraught with difficulties and dangers.[13] Therefore it is important to examine whether nuclear power should be regarded as a viable option for developing nations, at least in the immediate future.

Nuclear Technology Transfer

Nuclear power generation requires not only an exceptionally large commitment of economic and human resources, but a sophisticated industrial capability as well. The transfer of nuclear technology from one country to another is a massive and complex undertaking.

It is important to distinguish here between the transfer of nuclear technology from a supplier nation to a nation of comparable industrial capability, such as the transfer of light water reactor technology from the United States to European nations, and the transfer from an industrialized nation to developing nations such as Brazil or Iran, which do not yet possess the resources to support the technology. The transfer of nuclear technology to industrialized nations initially requires some assistance in the design, construction, operation, and maintenance of nuclear generating stations. But it will also frequently involve the transfer of manufacturing capability to local industries so that the nation can become reasonably independent in the support of nuclear technology.

The first step typically involves the sale of nuclear plants on a turnkey basis, that is, the supplier nation provides the entire plant and support technology. As a country's nuclear capabilities grow, the supplier nation may enter into technology and manufacturing exchange agreements to provide for two-way exchange of design information between local industries and firms in the supplier nation.[14] Experience has shown that industrialized nations are able to manufacture and eventually market nuclear equipment rapidly. Usually after a period of several years of operating under licenses from the supplier nation, they dissolve these ties and become an independent competitor in the international market. Examples of this nuclear self-sufficiency include France, West Germany, Italy, and Sweden.

The transfer of nuclear technology to developing nations is far more difficult.[15] Nuclear technology may not even be appropriate for many developing nations because of its capital-intensive nature and requirements for sophisticated industrial and manpower capability. Nevertheless many developing nations have rushed to acquire nuclear power technology, because it serves as a status symbol and because they perceive this energy source as a possible solution to the numerous problems created by growing fossil fuel shortages and price increases.

The highly sophisticated nuclear technology now being marketed by most supplier nations is frequently quite inappropriate for developing nations. For example, the smallest nuclear units that can be purchased (roughly 600 megawatts) are too large for the electric power systems of many developing nations. (In recognition of this incompatibility, the Babcock and Wilcox Corporation of the United States and Kraftwerk Union [KWU] of West Germany have announced plans to market plants in the 400 megawatt range.) The standards for the construction of nuclear power plants and their subsequent safe and reliable operation are frequently beyond the capabilities of developing nations. These nations quite commonly accept unchanged the Safety Analysis Report documents prepared by United States manufacturers for domestic reactor licensing, yet a number of problems may arise in siting nuclear power stations in developing nations that are not adequately addressed in these studies. For example, although electric power failures are extremely common in many countries, the plant frequently contains numerous systems that depend on the availability of off-site power. The highway and rail transportation networks of such countries are frequently inadequate for shipment of nuclear equipment or fuel materials. The degree of quality assurance needed for the construction and reliable operation of nuclear plants may exceed the capabilities of local contractors and industry. Furthermore, although safety review and research in the United States continually produces improvements in design and safety requirements, the design of plants sold to foreign nations is usually frozen at an early stage and will not benefit from this retrofit process.

The personnel requirements for planning, constructing, and operating nuclear power plants should not be underestimated. It is frequently difficult to educate enough engineers and technicians both to operate the plants and to regulate their performance. Usually the regulatory activities in developing countries are given second priority to plant operation and power requirements.

These considerations raise serious questions as to the suitability and safety of nuclear power technology transferred to developing nations. Such questions cannot be addressed by the suppliers themselves since they are involved in intense competition with one another in the international market. One must turn instead to supranational institutions such as the International Atomic Energy Agency. For example, the IAEA presently conducts a program of missions of qualified experts to review the safety of nuclear power facilities.[16] Such activities will become even more important as the worldwide commitment to nuclear power generation continues to grow.

Nuclear Power and Nuclear Weapons Proliferation

On May 18, 1974, the world was shocked by India's successful underground explosion of a nuclear device. This event signaled the beginning of an era in which nuclear weapons technology threatens to spread from a few nations of high technological capability, namely, the United States, the USSR, Great Britain, France, and China, to many nations including even poorer, underdeveloped countries such as India. The proliferation of nuclear weapons capability represents a serious threat to world peace. Although it may be several more years before another country explodes a nuclear device, at least a dozen countries have the technological capability to join the nuclear club any time they choose and another dozen will attain the capability to join this group within the next decade. Indeed, the Indian test demonstrated that even a poor country can accomplish the sophisticated task of successfully developing and testing an underground nuclear device if they are determined to do so.[17]

The proliferation of nuclear weapons capability is related to some degree to the spread of nuclear power technology. The major concern is not so much the nuclear power plants presently operating in some twenty nations, but rather the technologies for supplying and reprocessing the nuclear fuels utilized in such reactors. These technologies could be applied to the production of weapons-usable or strategic nuclear materials such as enriched uranium or plutonium.

The international proliferation of nuclear weapons capability by means of the diversion of such material from the nuclear fuel cycle is a far different and far more serious problem than the theft of the material by subnational groups such as terrorists. Of major concern is so-called *incipient proliferation*,[18] by which a nation might accomplish most nonnuclear aspects of weapons development with a modest and undetected effort. This action could lead to the overnight bomb scenario in which the nation would then have the capability to assemble a weapon rapidly when strategic nuclear material became available. In this manner many nations could move close in time and capability to possession of nuclear weapons.

The trend for nations to acquire an independent capability for nuclear fuel supply is particularly disturbing. For example, uranium enrichment technology is already fairly widely dispersed among nonnuclear weapons states including several European nations and South Africa. Furthermore a number of countries have developed independent spent fuel reprocessing capability including Belgium, West Germany, Japan, and India.

Ironically the United States has played a major role in stimulating such countries to acquire independent nuclear fuels production and reprocessing capability.[19] As long as such nations could depend on this country to supply uranium enrichment and spent fuel reprocessing services, there was little incentive to develop an independent capability. However, in 1974 the United States announced that, contrary to its previously stated policy, it was no longer in a position to provide enriched uranium to foreign markets. More recently the federal government's decision[20] to postpone indefinitely the development of spent fuel reprocessing technology has further eroded international confidence in American willingness or capability to supply a major fraction of the world's fuel cycle services. Little wonder then that industrialized nations already dependent on nuclear power decided to develop an independent nuclear fuel industry.

Although limiting the proliferation of nuclear weapons technology is certainly an admirable objective, it is also an extremely elusive goal that will require sophisticated and delicate action. Perhaps the major flaw in most present attempts to deal with the problem is an oversimplification of the relationship between nuclear power development and nuclear weapons capability. In this regard it should be noted that to date, nuclear power development has not played any significant role in nuclear weapons proliferation. Nuclear weapons programs have evolved entirely separately from nuclear power development.[21]

Several pitfalls can impede policies designed to halt weapons proliferation. Consider the accord in which Germany agreed to supply Brazil with an entire nuclear power industry, including nuclear power plants, enrichment facilities, and fuel fabrication and reprocessing facilities.[22] This represents the largest agreement ($5 billion) in history for an international transfer of nuclear technology. More significantly it also threatens to establish a commercial precedent that could contribute to the proliferation of weapons capability.

The basic contract calls for two to eight nuclear power plants provided by the West German consortium KWU. The Brazilian implementation of this technology will be conducted by the state nuclear agency NUCLEBRAS. The contract arranges for extensive uranium exploration and mining (with a clause for guaranteed delivery of 20 percent of the mined ore to German utilities). It also calls for the construction in Brazil of a jet-nozzle enrichment plant, a commercial fuel fabrication plant, and a spent fuel reprocessing plant. To quiet international criticism of this contract, Brazil has reluctantly agreed to international inspections to detect possible diversion of nuclear materials for weapons production that go beyond the nuclear safeguards normally required by the IAEA.

Actually the incentives for Brazil to look toward nuclear power as a future energy source are quite strong. Although Brazil is currently the world's seventh largest country in population and fifth largest in land area, it is extremely poor in fossil fuel reserves. Its hydroelectric resources are located far from major population centers. While Brazil has large proven deposits of thorium, the lack of available technology for thorium-fueled reactors has forced it to consider uranium-fueled light water reactors and to depend on extensive uranium exploration to develop unproven domestic uranium resources.

There is some doubt whether the nuclear energy transfer will meet Brazil's immediate needs for energy sources to offset its growing dependence on foreign oil imports. Furthermore the Brazilian technical community is concerned not only about the capability of their country to implement this massive nuclear commitment, but also about the purchase of an untried uranium enrichment process, the nozzle process. One skeptical Brazilian scientist describes the accord as "a trade of uranium that Brazil does not have for technology that Germany does not have."[23] The major concern of the international community is that this transfer of a complete nuclear fuel cycle technology will provide Brazil with the capability to develop nuclear weapons and set a dangerous precedent for future transfers of nuclear technology.

This latter concern has motivated the United States to apply strong diplomatic pressures to persuade both Germany and Brazil to cancel their agreement. Unfortunately this action has had just the opposite effect, for the United States failed to take note of the considerable internal opposition to the accord within the Brazilian technical community. By applying external diplomatic pressures, the United States has stimulated strong national pride that has strengthened the Brazilian government's resolve to proceed with the nuclear power contract, which it now perceives as a matter of national security. The interpretation of the agreement as a national security issue has effectively stifled all internal dissent.

Another example of the hazards involved in nuclear diplomacy is the case of South Africa, which has some of the largest uranium reserves in the world. Therefore South Africa is moving quite naturally toward nuclear power development. It also has developed a commercial enrichment process (the helikon process, which is closely related to the German nozzle process) along with spent fuel reprocessing capability. When the South African government approached United States industry for the purchase of two large light water reactor power plants, the United States government responded with strong concerns about the sale and even suggested that the required export licenses might be denied. Faced with this opposition, South Africa immediately turned to France to supply the plants. In this instance the response of the United States government not only led to a major loss in nuclear equipment sales (in excess of $2 billion), but it also removed any possibility of United States involvement in the establishment of safeguards over the South African nuclear power program.

It should be apparent that nuclear weapons proliferation is a political as well as a technical problem. Controls over nuclear technology, regardless of how stringent, are insufficient to prevent determined nations with a moderate degree of technological capability from developing nuclear weapons. Rather one must aim at removing those political pressures that push nations in that direction.[24]

Nuclear technology controls are not useless when coupled with appropriate diplomatic actions. International safeguards governing the transfer of nuclear materials are certainly useful in limiting proliferation. One can also make several technical modifications in the transfer of nuclear power and fuel cycle technology among nations. However, technical approaches by themselves are insufficient to prevent determined nations from developing weapons capability. The fundamental dilemma and challenge today is how to provide needed nuclear power technology while limiting the potential diversion of this technology for the production of nuclear weapons.

Technical Aspects of Nuclear Weapons Proliferation
The primary technical requirements for a nuclear weapons program include basic information on nuclear weapons design; skilled scientists, engineers, and technicians who can design and fabricate the device; production and assembly facilities; a variety of nonnuclear components including sophisticated conventional explosives and electronics; and most significantly, strategic nuclear materials can be used as the bomb-stuff in a fission explosive.[25] The basic concepts of fission chain reactions and explosive fission devices are now openly available in the technical literature. While many specific weapons parameters, dimensions, and materials are still classified or difficult to obtain, a reasonably competent scientific team should be able to design a crude fission device without an extensive research program. Any country with a moderate technological base will have a number of scientists and engineers with the necessary technical training in basic concepts. In fact, a considerable number of foreign students have been educated in nuclear energy release in American universities during the past two decades.

The key ingredient in a weapons development program is access to strategic nuclear materials such as highly enriched uranium or plutonium. The rush of the world community to adopt nuclear power has increased the ease with which fissile material can be obtained. The nuclear fuel cycle that supplies and reprocesses power reactor fuel can be adapted to the production of weapons materials.

The same enrichment facilities designed to produce low-enrichment uranium for power reactors could be used with some modification to produce highly enriched uranium suitable for nuclear weapons. This approach, although difficult and expensive, might be attractive to a nation that wished to make only a few weapons for diplomatic leverage, since such uranium can be used in relatively crude gun-type weapons.[26]

The far more likely and therefore sensitive point in the nuclear fuel cycle at which strategic nuclear materials can be obtained is in the reprocessing of spent reactor fuel. The fuel discharged from power reactors contains a significant amount of plutonium. For example, a 1,000 megawatt light water reactor discharges some 200 to 250 kilograms of plutonium each year in the form of spent fuel.[27] This material can be separated out by chemical methods and processed into a form suitable for weapons designs. In a similar manner the spent reactor fuel elements from thorium-fueled reactors can be reprocessed to extract uranium-233, which could also be used in weapons.

The technical key to the development of nuclear weapons capability, and therefore to the control and prevention of nuclear weapons proliferation, lies in the controls that are placed on strategic nuclear materials.

Enrichment. The most difficult path to nuclear weapons development is the acquisition of an independent capability to enrich uranium to the levels necessary for weapons applications (greater than 20 percent). Of course, such a clandestine enrichment operation would be capable of producing only a small amount of fissile material. Furthermore the technology of uranium enrichment, unlike that of spent fuel reprocessing, has been heavily classified, and much of the necessary information is not available in the open technical literature. Nevertheless highly enriched uranium does have the advantage that it requires a far less sophisticated weapons design than does plutonium. It therefore might present a more desirable option to a country determined to obtain a limited nuclear weapons capability.

There is a variety of enrichment processes that might be used in the production of weapons-grade uranium. The elaborate complexes required for uranium enrichment that utilize gaseous diffusion or ultracentrifuge processes are probably beyond the capability of most nations to construct or operate. However, less demanding techniques such as aerodynamic nozzle or thermal diffusion methods might present attractive alternatives. Furthermore the successful development of exotic new methods such as laser isotope separation with its single enrichment stage capability may change this picture dramatically and could have important implications for nuclear weapons proliferation.[28]

The old-fashioned gaseous diffusion plants are far less suitable for the production of strategic nuclear materials than the newer technologies,[29] since a plant designed to produce low-enrichment fuels would have the wrong separation unit sizes and flow rates to produce weapons grade material. The large number of stages required for appreciable enrichment would necessitate a major effort to convert a gaseous diffusion plant over from low to high enrichment. The same conclusion would hold for the aerodynamic nozzle separation method, although an adjustment of the blade settings in the nozzle devices would facilitate higher enrichment operation. (In both cases it would probably be easier to add a topping cycle based on electromagnetic separation methods to obtain higher enrichment, much as was done at Oak Ridge during the Manhattan Project.) By way of contrast it has been estimated that a gas centrifuge plant adequate to meet the needs of a small weapons program could probably be built at a cost of several tens of millions of dollars.[30]

Although this country has been extremely reluctant to export enrichment technology, other countries such as West Germany have agreed to provide this technology. Any export policies that affect enrichment technology must therefore take into account not simply the potential for diversion of strategic nuclear materials from a particular process, but also the hazards that failure to satisfy the demands of the international market may cause. For example, to refuse enrichment technology to a country may trigger a large-scale research effort to develop an independent enrichment capability.

Spent Fuel Reprocessing.   In contrast to the difficulty of separating isotopes of the same element, the separation of elements such as plutonium from uranium can be accomplished by relatively straightforward chemical methods. Plutonium occurs in several forms in the nuclear fuel cycle. When it is contained in a spent fuel element, it is associated with so much radioactivity that it is effectively inaccessible. However, after several years this radioactivity will decay to such a level that extraction by a sophisticated group would be conceivable. Of course, after the spent fuel has been reprocessed and the plutonium has been separated and processed into a mixed uranium-plutonium oxide form for recycling, it is far more accessible, although it would have to be separated from the uranium and converted into a metal, a task that is relatively difficult, but well within the technology of most nations.

The technology required to reprocess spent reactor fuel elements and recover plutonium was made public at international meetings during the mid-1950s, and a number of countries have acquired limited reprocessing capability. However, this particular approach to weapons development is not quite as straightforward as it might first appear because of the particular isotopic composition of power reactor grade plutonium. We have noted that the plutonium discharged from power reactors contains appreciable concentrations of the isotopes Pu-240 and Pu-241 because of the high fuel burnup in most power reactor designs. These isotopes are characterized by significant spontaneous neutron emission and radioactive decay heat. This can lead to predetonation of an explosive device unless some sophistication is used in its design.[31]

Although this particular feature of reactor grade plutonium presents a significant barrier to the small terrorist group engaged in amateur bomb building, a determined nation with competent scientific capability should be able to design an inefficient explosive weapon from even reactor grade plutonium. Although the plutonium discharged from power reactors is not "weapons grade," it is certainly "weapons usable." Furthermore, by appropriately tampering with the reactor fuel cycle (by changing the reload frequency for example), one can significantly reduce the percentage of the higher plutonium isotopes present in spent reactor fuel. For only a modest investment (tens of millions of dollars)[32] a determined nation could obtain "clean" Pu-239 by reprocessing low-burnup fuel discharged from natural uranium research reactors, since these reactors can be refueled on a continuous basis and operated at low power densities to avoid the production of the higher plutonium isotopes. The reprocessing of low-burnup fuel is a considerably easier task than the reprocessing of highly radioactive spent fuel from power reactors. India apparently tampered with the fuel cycle of a Canadian-supplied research reactor to produce fissile material for their first nuclear device. In fact every country that has produced a nuclear weapon has used plutonium produced in either a Hanford-type plutonium production reactor or a research reactor (with the possible exception of South Africa). To our knowledge no plutonium produced in a power reactor has ever been used to fabricate a nuclear weapon.

Hence we must conclude that any nation bent on obtaining strategic nuclear materials could do so by acquiring a modest spent fuel reprocessing capability and tampering with research reactor fuel cycles to produce a more suitable form of plutonium, although to do so would probably be tantamount to announcing a national intent to acquire nuclear weapons. In this sense the rapid spread of nuclear power technology makes a relatively modest contribution to nuclear weapons proliferation.

Nevertheless, a high degree of concern has been expressed about the transfer of spent power reactor fuel reprocessing technology to nonnuclear states. For example, when France agreed to supply such technology to Pakistan and South Korea, the United States applied strong diplomatic pressures that eventually led to the cancellation of the South Korean sale and a reconsideration of the sale to Pakistan.[33] The United States also applied pressure to Japan in an effort to slow the development of a commercial reprocessing facility at Tokai-Mura.

The United States government has called on the rest of the world to accept an indefinite postponement of spent fuel reprocessing.[34] It has backed up this request by canceling its own fuel reprocessing plans. This policy seems to have had little effect aside from stimulating strong international dissent.[35] Great Britain, France, Belgium, West Germany, and Japan all intend to proceed with national reprocessing plants. Except for Belgium whose reprocessing plant is presently being refurbished, all are currently separating pure plutonium. The incentives for reprocessing spent power reactor fuel stem from the 30 to 40 percent reduction in uranium feed requirements achieved by plutonium recycle in light water reactors and from the necessity of fuel reprocessing for plutonium-fueled fast breeder reactors. Those nations that have made strong commitments to fast breeder reactor development have made similarly strong commitments to develop spent fuel reprocessing technology.

Technical Fixes on the Nuclear Fuel Cycle.   A number of technical fixes have been proposed to modify the nuclear power reactor fuel cycle in such a way as to limit the proliferation of nuclear weapons capability.[36] These proposals attempt to modify the nuclear fuel cycle to minimize the presence of separated, or easily separable, weapons-usable plutonium. Here we should take care to distinguish between the relatively minor actions that would prove sufficient to thwart subnational theft of strategic nuclear materials, such as collocation of fuel reprocessing and fabrication, enhanced physical security measures, spiking of strategic nuclear materials with radioactive sources, and those major modifications to the fuel cycle required to impede significantly diversion of weapons-usable material by national governments.

Although we will review briefly the various alternative "proliferation-resistant" fuel cycles now being proposed, we fear that the proliferation problem has evolved far beyond the point where such technical solutions are likely to have any appreciable impact. There are now so many routes available to a determined nation that mere restrictions on the transfer of nuclear power and fuel cycle technology will have only a limited effect.

In a sense the present light water reactor fuel cycle already contains an inherent technical fix, since the high-burnup plutonium it produces is quite inappropriate for nuclear weapons. Although reactor grade plutonium can be used in the manufacture of crude nuclear devices, such weapons would have questionable strategic military value, although they might have symbolic value.

The most dramatic technical fix would be simply to phase out nuclear power development altogether. However, a world faced with impending exhaustion of its fluid fossil fuel resources is unlikely to accept such a severe proposal, so we will dismiss immediately this action as an unviable approach.

A less draconian, but probably no more acceptable, long-term variation on this theme is the stowaway or throwaway fuel cycle proposed by the United States.[37] This approach would avoid spent fuel reprocessing altogether and either store or dispose of spent power reactor fuel in its original form (zircaloy-clad uranium oxide fuel pellets). This approach, coupled with the decision to slow down fast breeder reactor development, has been adopted by the United States government to demonstrate our willingness to take the lead in removing strategic nuclear materials from the nuclear fuel cycle. This stance amounts to throwing out the baby with the bathwater. This approach not only "discards" roughly one-third of the energy content of light water reactor fuel, but it also eliminates the long term potential of breeder reactor fuel utilization. Quite naturally it is an unacceptable alternative for most nations facing more imminent energy shortages than the United States faces.

A somewhat less drastic alternative involves the coprocessing of spent reactor fuel.[38] Such a scheme simply avoids any separation of plutonium from the uranium extracted from spent fuel elements during the Purex process. The product would be a dilute solution of about 1 percent fissile plutonium in uranium. Although this particular approach would prevent terrorists from obtaining strategic nuclear materials, the chemical separation of plutonium from uranium in the absence of fission products is well within the capability of most national states. Furthermore coprocessing requires a significant increase in reprocessing and fuel fabrication capacity since much larger material volumes must be handled. This approach would also not be an option for breeder reactor fuel cycles that require fissile concentrations of 15 percent or higher. A modification of this scheme is being implemented by Japan under strong United States pressure in its Tokai-Mura reprocessing plant.

Yet another modification of the usual light water fuel cycle involves spiking the plutonium extracted by a reprocessing facility with high radioactivity so that it presents a radiological hazard to those who might attempt to divert it.[39] One could simply leave a certain fraction of the fission products in the product stream. However, considerable experience from early AEC programs indicates that the fabrication of recycle fuel containing fission products is likely to be prohibitively expensive for light water reactor applications. An alternative would be to attach large masses of radioactive materials to shipping containers of the fuel. But these options are geared more for inhibiting subnational theft of strategic materials than preventing diversion by national governments.

A much different approach involves the use of a tandem fuel cycle in which the spent fuel from light water reactors is recycled in other reactors. Heavy water reactors such as CANDU reactors can operate on the low fissile concentration in spent light water reactor fuel to extract additional energy. Reprocessing and plutonium separation would be unnecessary. One could either refabricate the light water reactor fuel into a form suitable for CANDU reactors, or redesign light water reactor fuel elements so that they could be used in both reactor types. Unfortunately the refabrication of radioactive spent fuel is likely to be extremely expensive.[40] Furthermore heavy water reactors are themselves well suited to the production of weapons-grade (low-burnup) plutonium since they can be refueled on a continuous basis. Hence the technical fix in this instance may present a greater proliferation danger than the original light water reactor fuel cycle.

A closely related approach is the use of the spectral shift reactor in which fuel utilization can be varied throughout fuel life. The moderator in this reactor is a mixture of D_2O and H_2O that can be varied to change the core characteristics from high conversion (high plutonium production) at the beginning of core life to high burnup (to deplete the plutonium) toward the end of core life. Such a reactor would extract considerably more energy from the fuel in a throwaway fuel cycle than would the conventional light water reactor design.

An alternative that has caught the fancy of the arms control community is a thorium fuel cycle[41] in which the uranium-233 produced by thorium conversion is recycled. Although pure U-233 is an excellent material for weapons fabrication, it can be separated out of spent fuel elements and denatured by mixing it with U-238 in proportions of 1 to 8. This presumably destroys the weapons capability of the fuel, since it would require isotope separation to increase the fissile concentration to a level high enough for weapons use. The buildup of the radioactive isotope U-232 during U-233 production also would make this fuel more difficult to handle.

Some plutonium would be produced by U-238 conversion, and some additional fissile feed would be needed to sustain such thorium-fueled reactors. So this approach usually envisions a thorium reactor power industry sustained by several high-security centers in which U-233 is separated out of spent fuel elements and in which plutonium-fueled fast breeder reactors produce the excess fissile material needed to fuel the thorium reactors. Plutonium is inevitably produced in all such denatured fuel cycles. Thorium fuel cycles reduce plutonium production to some extent, but they do not entirely eliminate it.

The thorium and denatured uranium alternative presents other difficulties. Only a modest isotope separation facility such as a garage full of simple gas centrifuges would be needed to reseparate out the U-233.[42] Since irradiated thorium passes through an intermediate stage, protactinium-233, with a rather long radioactive half-life of thirty days, on its way to being transmuted into U-233, it might be possible to tamper with the fuel cycle to separate out protactinium chemically before it decays into U-233. The major argument against the thorium fuel cycle is that there is presently little experience with thorium-fueled power reactors. There is presently no thorium fueling in commercial light water reactors, and there are presently no plans for its commercialization. To convert the present uranium-fueled reactor industry over to a thorium fuel cycle would take roughly two decades, which unfortunately is the most critical period for nuclear weapons proliferation.

None of these technical fixes is compatible with the development of a uranium/plutonium-fueled fast breeder reactor. However, an alternative scheme known as the CIVEX process (for civilian reprocessing) has recently been proposed.[43] This scheme takes advantage of the relative insensitivity of fast reactors to the presence of fission products, in sharp contrast to thermal reactors that would suffer from a significant degradation in performance. The CIVEX process would use automatic, remotely controlled equipment to process a product stream in which plutonium, uranium, and a significant fraction of fission products would never be separated. This reprocessed fuel would be so highly radioactive that it could not be used in nuclear weapons fabrication. The CIVEX reprocessing plant equipment and layout would make physically impossible any operator manipulation or process modification that could isolate weapons-usable plutonium. Much of the technology for a CIVEX operation has been developed for on-line fuel reprocessing in the EBR-II experimental breeder reactor project in Idaho. This proposal is currently under review by both the federal government and the nuclear power industry.

It is doubtful whether these alternative fuel cycles can halt or even significantly slow the proliferation of nuclear weapons capability, however. At best they can merely complicate the access to weapons-usable material in the fuel cycle. And it should be kept in mind that alternatives such as coprocessing, the tandem fuel cycle, and the thorium fuel cycle were examined during the early days of the nuclear power development program and discarded in favor of the present light water reactor fuel cycle. In many cases these technical fixes may create as many new proliferation problems as they are intended to solve. Perhaps the most attractive technical fix on limiting proliferation is the present light water reactor fuel cycle coupled with some minor modifications to limit the diversion of plutonium (see table 18).[44]

A fundamental flaw in all of these proliferation-resistant fuel cycles is that they fail to recognize the enormous variety of paths that a determined nation can take to obtain strategic nuclear materials (see table 19).[45] These range from a relatively modest program using simple natural uranium-fueled Hanford-type reactors to produce weapons-grade plutonium to advanced schemes of isotope separation or plutonium production based on technologies presently under development, such as laser isotope separation or fusion neutron sources. With this enormous number of options, it is unlikely that a nation would choose to divert a multibillion dollar power reactor industry to the production of reactor-grade plutonium for nuclear weapons purposes.

There is probably no practical way technically to prevent any nation from acquiring nuclear weapons. Technical and political considerations cannot be separated. Limiting proliferation must deal with the political incentives that lead a nation toward nuclear weapons development. Technical fixes such as denatured fuel cycles can diminish the threat of subnational theft and limit access to weapons-usable material, but there are no insurmountable technical

TABLE 18. A Formula to Reprocess without Proliferation

1.	Reprocess normally in plants in weapons states or in multinational plants
2.	Collocate plutonium fuel fabrication plants with reprocessing plants
3.	Briefly irradiate fuel elements after fabrication and before shipment
4.	Ship in spent fuel casks

Source: K. Cohen, "The Science and Science Fiction of Reprocessing and Proliferation" (Paper presented at the Nuclear Fuel Cycle Conference, Kansas City, Mo., 1977).

TABLE 19. Sources of Strategic Nuclear Materials

Ways Presently Available	Required Cost	Required Technology	Required Industry
Research reactor	small	small	small
Production reactor	medium	medium	medium
Power reactor	large	large	large
Gaseous diffusion	large	large	large
Centrifuge	medium	medium	medium
Nozzle devices	large	medium	large
Electromagnetic separation	medium	large	medium
Accelerator	medium	medium	medium

Future Sources

U-235 separation using
1. laser isotope separation
2. chemical exchange
3. jet membrane
Pu or U-233 production using
1. magnetic fusion hybrid (fusion/fission system)
2. inertial confinement fusion (laser, electron beam, or ion beam)
3. accelerator neutron sources

Source: C. Starr, *Nuc. News* 20:54 (1977).

barriers to constructing a nuclear explosive if a nation is determined to do so. Ultimately nonproliferation depends on political restraints.

International Safeguards against
Nuclear Weapons Proliferation
Although a variety of proposals[46] for restricting the spread of nuclear weapons technology received consideration during the two decades following World War II, the first major international agreement, the Treaty on the Nonproliferation of Nuclear Weapons was not adopted until 1968.[47] This treaty banned the acquisition of explosive nuclear devices by nonnuclear states and set up a program of international safeguards and inspections to detect unauthorized diversion of nuclear materials. The Nuclear Nonproliferation Treaty was accompanied by commitments to disarmament by both the United States and the USSR. The nuclear states also pledged to help signers of the treaty develop peaceful applications of atomic energy.

To date some one hundred countries have signed this treaty. But more significantly, such countries as Argentina and Brazil, which possess not only the technology but possibly also the inclination to develop nuclear weapons, have not signed the treaty. Neither has Israel nor South Africa, which may already have such weapons. Three nuclear states, France, China, and India, also have not signed the treaty. Moreover, the governments of Egypt, Pakistan, and twelve other signatories have continued to withhold ratification of the treaty.

The safeguards procedures established by the treaty are administered by the IAEA. However, such safeguards do not provide any physical control over nuclear materials. Instead they provide a warning if such material is diverted for an illicit purpose. The safeguards consist of accounting and sampling procedures to keep track of nuclear materials, tamperproof containment seals, video monitoring devices, and other physical barriers, along with occasional international inspection to verify the location of such materials.[48] The purpose of such safeguards is to deter the diversion of strategic nuclear materials by posing a risk of detection and providing a basis for international action against any violator. The safeguards are regarded as a "burglar alarm, but not a lock."[49]

India's successful explosion of a nuclear device in 1974 not only marked the beginning of an era in which nuclear weapons could be easily acquired by almost any determined country, but also signaled a failure of the Nuclear Nonproliferation Treaty. Its most significant drawback has been its failure to attract the participation of those very countries that are most likely to develop independent nuclear capability. Superficially one can point to a failure of the superpowers to live up to their pledges of significant nuclear weapons disarmament along with failure to provide significant assistance in the development of peaceful nuclear technology. But a more significant reason for the limited success of the treaty has been its failure to take account of the strong political pressures on nations to acquire nuclear weapons. Certainly the experience of India has indicated that a country's decision to acquire nuclear arms has little to do with the arms race among the superpowers. Rather they acquire arms for reasons such as fear of neighboring countries (in India's case, China) or the use of nuclear weapons as a regional status symbol. Certainly, too, there is the mistaken belief that security is conferred by possession of unsophisticated nuclear forces. Furthermore for many nations there are substantial psychological advantages in having nuclear weapons capability. Although the nuclear powers swiftly condemned India's action (and have since reacted once again to dissuade South Africa from testing a nuclear device in 1977), the majority of third world opinion has ranged from neutral tolerance to admiration.

Several other countries may soon follow India's example. This group includes countries of high technology such as Israel and South Africa (both of which probably already have some nuclear weapons capability), as well as Taiwan, Argentina, Brazil, Chile, Egypt, Iran, Pakistan, and Spain, which may feel external or internal political pressures to develop such weapons.

A Pragmatic Approach to Limiting Proliferation

So what is to be done? How can the spread of nuclear weapons be halted? Certainly an effort should be made to restrict the international spread of strategic nuclear materials. One suggestion which has received considerable attention in the United States involves restrictions on the export of nuclear power technology. To the extent that this approach would restrict export of nuclear power plants themselves, it would be self-defeating. The incentive in many parts of the world to acquire such technology has become overwhelming due to the pressures of ever-increasing population and the depletion of existing reserves of fossil fuels. Furthermore a number of foreign nuclear equipment manufacturers are only too anxious to supply nuclear power plants to any potential customer. Consequently the number of reactors exported by United States companies has decreased dramatically in recent years, and we are in danger of losing our foreign market for nuclear equipment. Certainly we cannot hope to control the nuclear materials produced in the reactors of foreign manufacturers. Furthermore, the light water reactors exported by this country are perhaps the safest from the point of view of nuclear weapons proliferation, since the type of plutonium they produce makes inefficient explosive devices. By refusing to export a technology that is clearly capable of easing the problems of fuel shortages in many parts of the world, the United States is pushing such nations closer to the brink of international conflict.

A more rational approach would be to restrict the export of nuclear fuel cycle technology rather than power plants themselves. For example, greater care should be taken in the export of plutonium reprocessing technology. Representatives from fifteen nations have been meeting for the past several years in London to draw up guidelines for nuclear technology transfer. The London Nuclear Suppliers Group has agreed that any export reprocessing technology should be accompanied by the most stringent controls, preferably administered by an international organization such as the IAEA.

Other proposals call for no export of reprocessing technology. One possible alternative would be to lease reactor fuel to foreign nations outright and then reclaim the spent fuel for reprocessing or disposal in this country. That is, we would sell only energy to foreign nations, not fuel. An extension of this approach is to abandon the effort to separate plutonium from spent fuel elements (plutonium recycle) to prevent the spread of nuclear weapons-usable material. Under a leasing scheme the spent fuel elements would be returned to this country and stored without reprocessing.

To this end the United States has modified its own nuclear power program by deferring indefinitely the reprocessing of spent fuel and plutonium recycle and by slowing the development of the fast breeder reactor. The rationale behind this policy is that it will demonstrate the federal government's belief that plutonium reprocessing (and perhaps even the breeder reactor) is unnecessary. This vivid demonstration of the United States's commitment to nonproliferation will presumably set an example for other nations.

The resentment and dissent stirred in other nations by the United States's policy make it quite apparent that the policy is doomed to failure. First, the decision to defer plutonium recycle and breeder reactor development was based on optimistic estimates of domestic uranium reserves,[50] as well as reliance on conventional fossil fuels (primarily coal), conservation, and the development of new technologies such as solar power. Since most other nations of the world are blessed with neither the domestic energy reserves nor the faith in unproven technologies of the United States government, they are understandably reluctant to commit themselves to an inefficient throwaway or once-through fuel cycle for light water reactors. They recognize that the longer the United States postpones the use of plutonium as fuel, the faster it will use up the world's supply of uranium. Other nations are alarmed by this country's increasing consumption of world petroleum resources. The United States's decision to use its own nuclear resources inefficiently is hardly likely to persuade resource-poor countries to follow our moral and ethical leadership.

Other supplier nations suspect that the present United States policy is merely a thinly veiled attempt to slow the development of a competitive nuclear power industry in these countries. This attitude is particularly evident in France and West Germany, which have surpassed the United States in the development of advanced nuclear technology. The unilateral decision by the United States to halt development of spent fuel reprocessing represents in a sense a failure to live up to its obligations under the Nuclear Nonproliferation Treaty, which requires it to cooperate in international nuclear power development.

The present United States nuclear policy is contributing to rather than slowing the spread of nuclear weapons capability, since it forces nations to develop independent spent fuel reprocessing capability. The decision has also damaged significantly its own domestic nuclear power industry and has raised the fears of electric utilities that they may be left holding spent fuel for a long period of time. The industry also fears possible shortages of uranium over the next several decades if the administration's projections of domestic resources prove overly optimistic.

Any effective nonproliferation policy must recognize the legitimate security concerns of nations that might impel them to acquire nuclear weapons. And, however frightening it may be, we must admit that the nuclear genie is out of the bottle. Many countries have nuclear weapons capability already, whether or not they choose to use it, since they have significant nuclear power industries and reprocessing or enrichment facilities. The natural evolution of nuclear technology alone makes it extremely difficult to prevent a nation from acquiring nuclear weapons. The illusion that restricting commercial nuclear power development will somehow solve the proliferation problem is extremely dangerous. Such limitations would almost certainly be counterproductive. In fact, one should stress instead that tenuous relation between commercial nuclear power development and nuclear weapons proliferation to encourage a realistic assessment of the more significant political factors involved in a nation's decision to develop nuclear weapons.

The only realistic approach to preventing the further spread of nuclear weapons is to demonstrate clearly to nonnuclear states that it is not in their best interests to acquire such capability. The strongest pressures on nations to acquire nuclear weapons will be caused by their impending shortages of the basic resources, food and energy, in a time of exploding populations. A starving world is a dangerous world. We must reduce those pressures by providing nations with the technology and assistance they need to meet their own demands for energy, food, and industrial development. For those nations that choose to acquire nuclear power technology, we must provide suitable assistance. The United States can play an important role in limiting proliferation by assuring an adequate supply of nuclear fuel, thereby deterring foreign development of sensitive fuel processing facilities. The United States should resume its development of nuclear fuel reprocessing technology and breeder reactors to stretch existing uranium reserves, and it must develop and implement an energy policy that will reduce its massive consumption of the world's petroleum resources.

It is important to realize that nuclear weapons proliferation is beyond the control of the United States or any individual nation. The United States must forego the temptation of unilateral action. An effective nonproliferation strategy demands the active participation and interaction of the entire world community.

6

Advanced Forms of Nuclear Power

Nuclear power generation has evolved into a mature technology capable of meeting a significant fraction of the world's needs for electric energy. But nuclear energy generation still faces serious technical, social, and political problems. These problems will almost certainly limit its suitability or desirability for massive deployment. Nuclear power is still very much a technology on trial.

The principal form of nuclear power generation in use today relies on light water reactors, which are quite inefficient in utilizing uranium fuel. This particular form of nuclear power will experience only a relatively short period of viability lasting perhaps half a century because of its limited resource base. Therefore, if nuclear power is to play a significant role in our long term future, we must move now to develop advanced nuclear technologies that draw on far larger resources. There are two leading contenders as future nuclear power sources: breeder fission reactors and controlled thermonuclear fusion.

The Breeder Reactor

The present generation of light water fission reactors extract only about 1 percent of the energy contained in our uranium ore resources. This corresponds essentially to the U-235 component of natural uranium. Therefore the effective resource base for light water reactors is quite limited—of about the same magnitude as our domestic resources of petroleum and natural gas. Although these resources should fuel several hundred light water reactors for their operating lifetimes, use of this type of reactor must be regarded as an interim and relatively short-term energy option.

Of course, all reactors are partially fueled with fertile materials such as U-238 and Th-232. During operation these materials capture neutrons from the chain reaction and are transmuted into fissile fuels such as plutonium or U-233. Hence there is strong incentive to design a power reactor that will maximize this conversion process and make the most efficient use of our limited uranium and thorium resources.

Basic Principles of Breeder Reactors
While the goal is to design a reactor that maximizes conversion of fertile to fissile material, there is a conflict in design objectives here. This conversion process is accomplished most effectively with high-energy or fast neutrons. But the fission chain reaction operates most easily—that is, with smallest critical mass—with slow or thermal neutrons. For this reason the first generation of power reactors minimized fuel inventory requirements by using light materials such as water or carbon to moderate or slow down the fission neutrons. Unfortunately these reactors are rather ineffective at converting fertile into fissile material. For example, light water reactors are characterized by a conversion ratio of production of fissile material to destruction of fissile material of about 0.5 to 0.6. More advanced reactor designs achieve larger conversion ratios and therefore utilize fuel more efficiently. One such advanced converter reactor is the high-temperature gas-cooled reactor that achieves a conversion ratio of 0.8. This yields a fuel requirement of approximately half that of the light water reactor.

The long-range goal in reactor design is to achieve a conversion ratio greater than 1. More fuel would be produced by conversion during reactor operation than would be consumed by fission. Such a breeder reactor would utilize as much as 70 percent of the energy content of our uranium resources by converting U-238 into plutonium as it produces power. The most common breeder reactor design uses fast neutrons to sustain the chain reaction and achieves a conversion or breeding ratio of 1.3 to 1.5. In other words, the fast breeder reactor produces from 30 to 50 percent more fissile fuel than it consumes.

Although breeder reactors could be fueled with either U-238 or Th-232, the most favorable conditions for breeding occur with a U-238/Pu-239 breeding cycle. In such a breeder the reactor core is loaded with a mixture of 85 percent uranium and 15 percent plutonium. This core is then surrounded by a blanket of natural or depleted uranium (U-238), which can capture neutrons leaking from the chain reaction in the core and convert still more uranium into plutonium. In a sense the fissile material that actually sustains the chain reaction in the breeder can be regarded as a catalyst in the transmutation process of uranium to plutonium.

Although the breeder reactor produces more plutonium than it consumes, it must still be shut down periodically and refueled. After the fuel assemblies in the core and blanket have undergone prolonged neutron irradiation, they are removed and reprocessed to remove fission products and recover plutonium. Since more plutonium is recovered than is needed to refuel the reactor, excess plutonium can be set aside to fuel other reactors. The time required for a breeder reactor to produce enough excess plutonium to fuel a second reactor is referred to as the *doubling time*. Most breeder reactor designs strive for doubling times of ten to twenty years so that they can produce plutonium at a rate that matches the increase in the demand for electric energy. The excess plutonium production can easily be tailored to the demand so that a stock of excess plutonium would never be accumulated.

The fuel resources available to the breeder reactor are enormous since they include common fertile materials such as uranium-238 and thorium-232. Since breeder power reactors can extract over fifty times more energy from natural uranium ores than light water reactors, the reserves of high-grade uranium and thorium ores are sufficient to fuel a fast breeder economy for thousands of years. The low fuel costs of breeder reactor operation also would make it economically attractive to recover even very low-concentration ores, thereby expanding the effective resource base enormously. We have already stockpiled over 200,000 tons of U-238 as the tails products of uranium enrichment plants used to fuel the light water reactor industry, and sufficient plutonium has been produced in light water reactors to provide the initial fuel charge for several breeder reactors. By the year 2000 light water reactors will have produced some 700,000 kilograms of plutonium, enough for about three hundred fast breeder reactor cores.

Breeder Reactor Design
Fast Breeder Reactors. Most effort has been directed toward developing the liquid-metal-cooled fast breeder reactor, (LMFBR) which uses liquid sodium as a primary coolant.[1] The relatively large atomic weight of sodium minimizes the slowing down of fast neutrons to facilitate a chain reaction. However, since fast neutrons are less likely than slow neutrons to induce fission reactions, the core of a LMFBR must be compact and contain highly concentrated fuel. A primary coolant with excellent heat transfer properties is needed to accommodate the necessarily high reactor core power densities (about 400 megawatts per cubic meter). Sodium not only has excellent heat transfer properties, but also a high boiling point of 882°C

that allows the primary loop to operate at atmospheric pressures while still achieving high coolant temperatures.

Sodium tends to become radioactive as it is exposed to the intense neutron flux in the reactor core. Therefore LMFBR nuclear steam supply systems include an intermediate sodium coolant loop to isolate the steam generator from this radioactivity. Since sodium and water react together quite violently and produce hydrogen, the design of the steam generators is particularly critical. Most United States designs utilize a "loop" coolant system similar to that of light water reactors; the reactor core is contained in a reactor vessel, and the coolant is piped to heat exchangers outside of the vessel. European designs favor a "pot" design in which the reactor core, heat exchangers, and primary sodium pumps are all contained within a single vessel. The pot design is intended to minimize any loss of integrity in the primary sodium coolant circuit by avoiding the possibility of pipe breaks.

Since the sodium coolant loops can be operated at a somewhat higher temperature (545°C) than the light water coolant of conventional reactors, LMFBR power plants will exhibit a higher thermal efficiency than conventional plants (40 percent as compared with 33 percent). Liquid sodium also is less corrosive than water in its action on the steel components of the reactor and the coolant system. Corrosion embrittlement in liquid-metal-cooled reactors is not the serious problem that it is in water-cooled reactors.

A second fast breeder reactor design is the gas-cooled fast reactor (GCFR) that uses high-pressure helium as a primary coolant. The GCFR is essentially a marriage between LMFBR and HTGR technology. The core design is similar to that of the LMFBR and uses mixed oxide (UO_2/PuO_2) fuel clad in stainless steel. Surface roughing of the fuel element is used to increase the heat transfer to the high-pressure helium coolant that is pumped through the core at high speeds. The low density of helium leads to somewhat faster neutrons and therefore a somewhat larger breeding ratio than that of the LMFBR (1.5 compared with 1.3). This advantage is offset somewhat by the problem of residual heat removal in the event of system depressurization. Because of its high core power density and gas coolant, a temporary loss of core heat removal in the GCFR is tolerable for only several seconds before significant fuel melting occurs. Elaborate auxiliary heat removal systems are required as engineered safeguards for this reactor design.

Breeder Reactor Safety. One possible reason for public reluctance to support development of the fast breeder reactor involves the perception that such reactors are inherently more dangerous than light water reactors. The term *fast* conveys an impression that breeder reactors might be harder to control. Actually fast reactors do exhibit a more rapid response to reactivity variations, but they can be designed with strong negative feedback mechanisms that cancel control problems.

Rather, the primary concern arises because the fuel in a fast breeder reactor is not arranged in its most reactive configuration in order to facilitate cooling and fuel handling. Therefore an accident could occur in which the fuel would move into a more reactive configuration, say by melting, and this could lead to a supercritical chain reaction causing significant energy release. That is, the significantly higher concentration of fissile material in a fast reactor leaves open the possibility of a nuclear explosion, a possibility that is clearly absent in light water reactors. Furthermore, the presence of a much higher plutonium concentration in the core, coupled with the possibility of fuel vaporization suggests that the maximum credible accident in an LMFBR could be as much as an order of magnitude more serious than a loss of coolant accident in a light water reactor. Although most nuclear reactor engineers believe that such a supercritical reassembly of fuel is highly improbable, it has not been possible to confirm this postulate fully.[2] Therefore a hypothetical core disruptive accident continues to dominate fast reactor safety research. Fortunately, even if such a reassembly were to occur, the energy release could easily be contained in the reactor pressure vessel, as verified by tests on scale models. Furthermore fast reactor designs contain numerous protective systems to prevent core disruptive accidents in the first place.

Other aspects of fast reactor safety are similar to considerations involved in water-cooled reactors. There are even some clear safety advantages in the LMFBR design. For example, the loss of coolant accident is mitigated by the superior heat transfer characteristics of liquid sodium and the fact that the primary coolant system can be operated at atmospheric pressures. The large mass of sodium in the primary coolant system and the maintenance of the sodium coolant far below its boiling point provide a large thermal inertia. The possibility of a loss of coolant accident is minimized even further in pot-type designs. Some difficulties in working with liquid metals include sodium fires and sodium-water interactions, but these difficulties can be overcome by proper design. In summary there is no fundamental reason why fast breeder reactors should be any less safe than light water reactors, although it may be difficult to convince the public of this.

Environmental Impact. The environmental impact of a breeder power plant will be quite similar to that of more conventional nuclear power plants. Since it will operate at somewhat higher temperatures and therefore greater thermodynamic efficiency, it actually will discharge less waste heat. The reduced fueling requirements of the breeder are also an environmental plus, since they reduce mining and milling requirements. In fact there would be no need for further uranium mining until the depleted uranium tails stored at the isotope enrichment plants have been used (several hundred years supply). The breeder will produce an amount of radioactive waste comparable to that of light water reactors, although the higher fuel burnups will imply somewhat higher actinide concentrations.

The major concern is directed toward the fuel cycle of the breeder reactor since this involves a larger amount of plutonium than does that of the light water reactor. This poses a problem in attempts to control nuclear weapons proliferation.[3] Although a similar problem arises in the recycling of plutonium in light water reactors, the larger amounts of plutonium and the necessity for spent fuel reprocessing makes this problem more severe in breeder reactors.

Thermal Breeder Reactors. Although it is easiest to achieve a net breeding of fissile material in a fast neutron spectrum, breeding is marginally possible in a thermal reactor fueled with Th-232 and U-233. That is, the number of neutrons emitted per neutron absorbed, η, is slightly greater than 2 for thermal neutrons incident on U-233. Since this breeder reactor would require only about one third of the fissile inventory of a fast breeder design, it has generated great interest.[4]

However, since the breeding margin is so small for thermal neutrons, a thermal breeder design is dominated by the desire to minimize neutron losses by leakage or absorption. Neutron leakage can be minimized by designing a large core and utilizing a blanket of fertile material around the core. The problem of minimizing absorption is a bit more difficult. One must first reduce neutron losses due to fission product absorption by removing fission products from the reactor as quickly as possible. An even more serious problem involves the sequence of radioactive decays initiated by a neutron capture in fertile Th-232. An intermediate decay product, protactinium-233 (Pa-233), has a significant neutron capture probability. Neutron capture by this nuclide not only removes a neutron from the chain reaction, but it also eliminates a Pa-233 nucleus that would later decay to U-233. Hence all thermal breeder reactor designs go to great lengths to minimize neutron absorption in this isotope.

In solid-fuel reactors the only option is to place the Pa-233 in a lower neutron flux environment than U-233. This is achieved by adopting a "seed-blanket" design with regions of fissile U-233 dispersed in a background of fertile Th-232. Such a core was fitted into the Shippingport pressurized water reactor and began operation in 1977. To minimize neutron absorption in this light water breeder reactor (LWBR), control is achieved by moving the fuel seeds in and out of the core to adjust leakage.

A more dramatic departure from conventional reactor design involves the use of fluid fuels, which are continuously extracted from the core and processed to separate out Pa-233 and fission products to reduce neutron absorption. Most attention has been directed at the molten salt breeder reactor (MSBR) in which the fuel is a molten salt mixture, LiF-BeF_2-ThF_4-UF_4. The core is moderated and reflected with graphite. The fluid fuel is piped into the reflected core of the reactor where it becomes critical and generates fission power. Then the heated fuel flows through a heat exchanger to transfer the fission heat to an intermediate coolant loop, and eventually to a steam generator. Part of the fluid fuel is continually bled off the primary loop and processed to remove fission products and Pa-233. Although a prototype molten salt breeder reactor has been designed and built, this program is not receiving active attention at the present time.

Breeder Reactor Development
The fast breeder reactor occupies a rather unique position among long-range energy options since its scientific and technical feasibility and to some extent its commercial viability have already been established. It stands in sharp contrast to other long-term options such as solar power or nuclear fusion. Fast breeder reactors have been built and operated in this country and abroad for the past two decades.

The fundamental concept of breeding was first demonstrated in 1946 at Los Alamos in a small reactor experiment known as Clementine. A subsequent program developed the Experimental Breeder Reactor I (EBR-I), which was the first nuclear reactor to produce electricity in 1951. Other follow-on prototype fast breeder reactors were constructed during the ensuing two decades. Of particular significance has been the evolution of the breeder reactor development program from a series of small experimental facilities, to much larger prototype power reactors, and eventually to commercial demonstration plants. France, the United Kingdom, and the Soviet Union are presently operating such demonstration plants. These nations also have commercial-scale breeder reactors under construction or in the advanced design stage.

The most successful breeder reactor program has been that conducted by France. This program has evolved from the Rapsodie experimental research reactor commissioned in 1967 to the Phenix demonstration breeder reactor power plant that has been in operation since 1974. During its first two years of operation the 250 megawatt Phenix achieved the highest reliability of any power plant in the world. This experience has stimulated a commitment to the construction of a commercial prototype breeder reactor, the Superphenix, at Creys-Malville in the upper Rhone valley.[5] This 1,200 megawatt plant, scheduled for commissioning in 1983, will supply the power grids of France, Italy, and Germany. The Superphenix plant is intended only as a commercial prototype, therefore its capital costs will be significantly higher than those of a comparable-size light water reactor. Even so, the projected costs of electricity from this plant are in the same range as those of electricity produced by oil-fired power stations. The French government expects the first commercial orders for breeder reactors based on the Superphenix design to be placed in 1981–82. Breeder reactors will account for roughly 25 percent of France's installed electric capacity by the year 2000.

Other nations have significant fast breeder reactor development programs. Both the United Kingdom and the Soviet Union have been operating breeder reactor demonstration plants for several years. West Germany and Japan are scheduled to bring similar plants on-line by the early 1980s. All of these nations have strong intentions to deploy commercially viable breeder reactors by 1990 (see fig. 20).

The United States breeder reactor program stands in sharp contrast to the programs of other nations.[6] Although the United States was the early leader in breeder reactor development, the past decade has seen a significant erosion in its position. If we measure the success of the breeder program in terms of the timetable to achieve a demonstration breeder reactor, then the United States is presently some ten years behind the French, the British, and the Russians, since a United States demonstration plant is not expected to begin operation until the middle to late 1980s. This lag in breeder reactor development involves a host of factors including past technical decisions, political considerations related to licensing and funding of breeder reactors, the resolution of policy questions associated with the nuclear fuel cycle, and the willingness of the American public to proceed with the development of this energy alternative. The last factor is of particular importance, since the breeder reactor program is currently projected to cost more than $10 billion in public

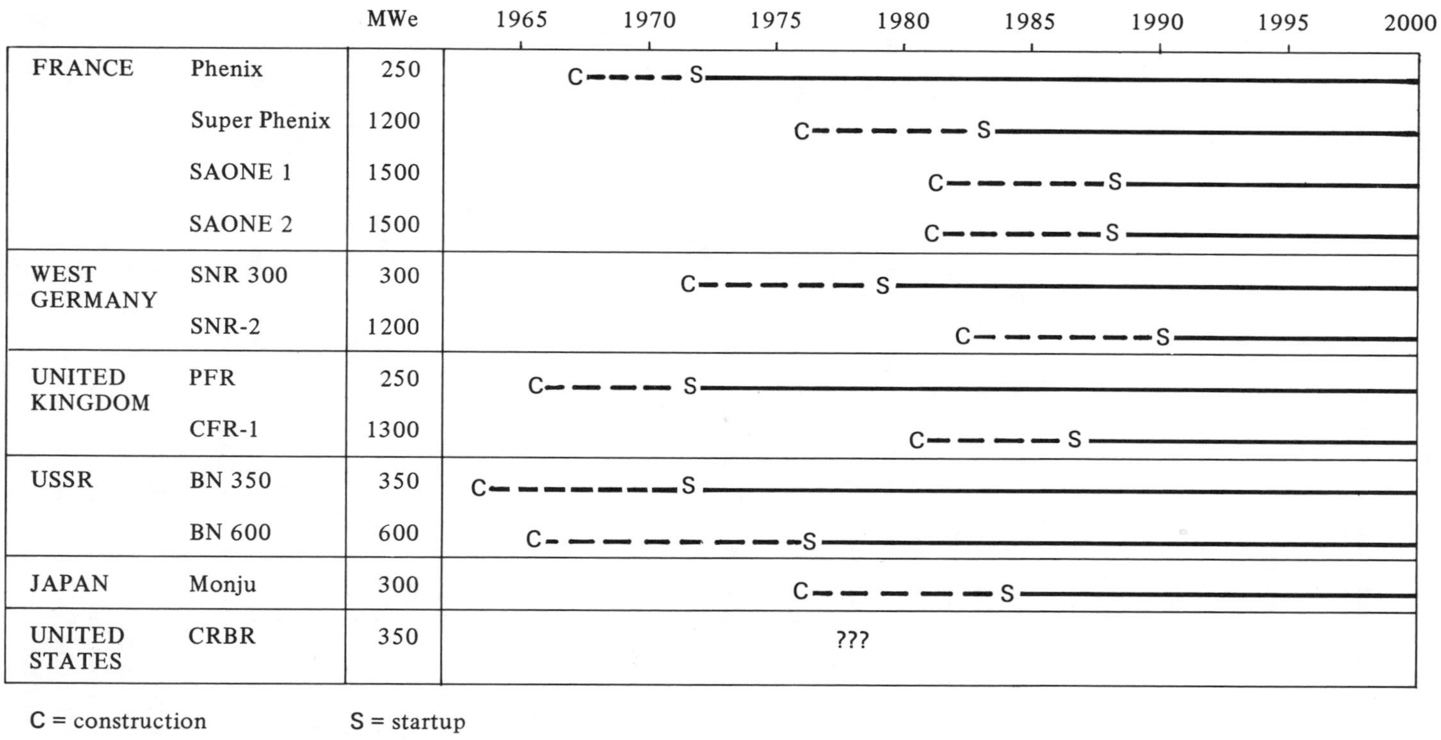

		MWe	Timeline (1965–2000)
FRANCE	Phenix	250	C–––S (≈1968 → 1973)
	Super Phenix	1200	C–––S (≈1976 → 1983)
	SAONE 1	1500	C–––S (≈1982 → 1988)
	SAONE 2	1500	C–––S (≈1982 → 1989)
WEST GERMANY	SNR 300	300	C–––S (≈1973 → 1980)
	SNR-2	1200	C–––S (≈1983 → 1989)
UNITED KINGDOM	PFR	250	C–––S (≈1967 → 1974)
	CFR-1	1300	C–––S (≈1980 → 1987)
USSR	BN 350	350	C–––S (≈1965 → 1972)
	BN 600	600	C–––S (≈1967 → 1976)
JAPAN	Monju	300	C–––S (≈1977 → 1986)
UNITED STATES	CRBR	350	???

C = construction   S = startup

Fig. 20. Timetable for fast breeder reactor development.

funds in contrast to the $3 billion invested in light water reactor development.

The United States attempted to demonstrate at an early stage the suitability of breeder reactors for commercial power generation with the construction and operation of the Enrico Fermi reactor near Detroit, Michigan, in 1966. This fast breeder reactor was built by a consortium of electric utilities under the leadership of Detroit Edison with assistance from the AEC. But it was considerably ahead of its time. Operating difficulties, including an accident in 1967 involving partial fuel melting, coupled with administrative decisions within the AEC to divert funding from the Fermi project to government-operated breeder reactors (the Experimental Breeder Reactor II, for example) led to its decommissioning in 1972.

A major change in the United States breeder reactor program occurred in the mid-1960s. Priorities were shifted from the development of demonstration or prototype plants to the building of a broad-based program in breeder reactor technology. An effort was made to attract the broadest possible involvement from private industry. The failure of this approach was vividly emphasized by the next large breeder reactor project, the Fast Flux Test Facility. This breeder reactor, which was designed to test fuel elements, was originally scheduled to go into operation in 1974 at a cost of $87.5 million. Program difficulties delayed its operation until 1980, and costs soared to $1.2 billion.

Even more dramatic has been the failure of the United States's demonstration project near Knoxville, Tennessee, the Clinch River Breeder Reactor Plant,[7] which was a reactor roughly analogous to the French Phenix plant (see fig. 21). Clinch River was originally scheduled for operation in the mid-1970s. Through a series of administrative shuffles and design changes, the Clinch River project has suffered many delays, and its estimated costs have soared to almost $2 billion. There are many engineers who feel that the present Clinch River design is already outdated and that the United States should bypass this project and move directly toward a breeder development program similar to those of France and Germany. More significant is the concern of the Carter administration that breeder reactor technology could accelerate the worldwide spread of nuclear weapons capability. Hence the Clinch River project has been deferred indefinitely to allow for the study of alternative breeder reactor concepts such as the light water breeder reactor and the thorium-fueled fast breeder reactor.

Roughly $4 billion has been invested over the past three decades in breeder reactor development in the United States, and an

Fig. 21. An artist's conception of the Clinch River Breeder Reactor Project. (*Courtesy of the United States Department of Energy.*)

additional $10 billion will be required for the development of a commercial prototype breeder reactor. The French and British programs have managed to proceed at only a fraction of this cost. For example, the French Superphenix plant will cost less than $1 billion.

The Carter administration has made every effort to curtail breeder reactor development, both in this country and abroad, because of concerns about the impact of this technology on efforts to control the spread of nuclear weapons. We have examined the effects of this policy in chapter 5.

The overwhelming motivation for developing and deploying the breeder reactor involves its ability to make far more efficient use of our resources of uranium and thorium than do the present generation of light water reactors. Therefore a critical factor in the urgency with which one views breeder reactor development is the estimate of our domestic resources.[8] If these resources are sufficient to fuel light water reactors until well into the next century, then a vigorous breeder program may be premature. However, if this resource base has been overestimated, then breeder reactor development should be given the highest priority.

There is no question that the present fast breeder reactor designs will successfully generate electric power. Rather, the real question is how fast this reactor should be developed. The high priority of breeder reactor development programs abroad is due in part to the desire to achieve greater independence in energy supply. This particular motivation may carry less weight in the United States, which has been blessed with considerable resources of coal and uranium. Nevertheless, a delay in breeder reactor development may entail a significant risk, for our domestic uranium resources may prove insufficient to fuel light water reactors beyond the turn of the century, and the generation of large amounts of energy from coal combustion may prove environmentally unacceptable. Other alternative energy sources may prove to be prohibitively expensive or unviable. Then the lack of a breeder reactor as a viable alternative could prove quite serious.

We should distinguish here between two separate decisions that must be made regarding the breeder reactor.[9] First, there is the decision to develop this reactor type, a decision that will involve a commitment of perhaps $10 billion in research and development. The second decision involves the commercial deployment of the breeder on a large scale. When the development cost of the breeder is interpreted as an insurance premium paid to cover the nation against the possibility of limited uranium resources and the inability to develop alternative energy sources, then $10 billion may be a small cost indeed. This amount presently represents about two months worth of imported petroleum. Fortunately, the decision to commit to a commercial breeder reactor economy need not be made for some time, provided we have developed the technology to give us this option.

Controlled Thermonuclear Fusion

The most glamorous of our long-range energy possibilities is *controlled thermonuclear fusion*.[10] This process involves a totally different kind of nuclear reaction than the fission reactions that drive today's nuclear power plants. Instead of splitting heavy atomic nuclei, the nuclei of light elements are fused together at enormous temperatures to produce energy. Nuclear fusion can be looked on as the most primitive form of solar power, since it is the energy source of the stars, and of our sun in particular. The awesome potential of this energy source was demonstrated by the development of thermonuclear fusion weapons—the hydrogen bomb—in the early 1950s. Since that time, proponents of fusion power have predicted that this nuclear process would someday provide man with a safe, clean, and abundant source of energy.

Nuclear fusion reactions are essentially the opposite of fission reactions. They involve the combining or fusing of light isotopes to generate more tightly bound, heavier isotopes, releasing energy in the process. An example of such a reaction is the one that occurs between the two heavier isotopes of hydrogen, deuterium and tritium:

$$\text{deuterium} + \text{tritium} \rightarrow \text{helium} + \text{neutron} + \text{energy}$$

This fusion reaction releases energy that is carried off by the reaction products, helium (an alpha particle) and a neutron. The potential of such reactions for generating enormous energies is evident. We need only look at the sun or any star to see a massive example of fusion energy release.

Unfortunately a rather major stumbling block that must be overcome before fusion reactions can occur. The light nuclei that must fuse together are both positively charged and repel one another quite strongly. To overcome this repulsion, we must slam the two nuclei together at very high velocities. One way of doing this is to take a mixture of deuterium and tritium and heat it to such a high temperature that the velocities of thermal motion of the nuclei overcome charge repulsion and initiate the fusion reaction. Such a scheme is referred to as a *thermonuclear fusion* reaction. The temperature required is quite high, roughly 100 million degrees. In fact, the interior of the sun is at just such enormous temperatures. Until quite recently man had imitated the sun in only a rather violent fashion by using a nuclear fission bomb to create temperatures high enough to ignite the fusion reaction in the hydrogen bomb.

Now we hope to be able to produce such high temperatures and initiate controlled thermonuclear reactions (CTR) in a controlled way. It is not too difficult to heat a gas to these temperatures, but it is quite difficult to contain the hot fuel long enough to get an appreciable amount of fusion energy out. The enormous force of solar gravity accomplishes this in the sun. Scientists on earth have taken a necessarily different approach by noting that such high temperature gases become ionized. Although no physical container could withstand the enormous temperatures of a fusion reaction, scientists have demonstrated that the ionized fusion fuel, or *plasma* as it is called, can be confined by a magnetic field. Although no one has been able to confine a high-temperature plasma in a magnetic field long enough to get appreciable fusion energy release yet, there is every hope that the scientific feasibility of achieving a controlled fusion reaction will be demonstrated within the next decade.

There may be another way to trigger the fusion reaction. Rather than confining the plasma, one might heat the plasma fuel so rapidly that its own inertia holds it together long enough for it to ignite and burn in a thermonuclear reaction. This *inertial confinement* approach is the basis of thermonuclear weapons, of course. To accomplish this in a controlled fashion, one must heat tiny pellets of fuel to fusion temperatures in roughly one billionth of a second. This can be accomplished by focusing the energy in a powerful pulsed laser beam on a fuel pellet of deuterium and tritium in such a way that the pellet is compressed and heated to thermonuclear burn conditions. The pellet then ignites and burns in a rapid thermonuclear reaction, in effect, a microthermonuclear explosion. The heat produced in this explosion can be used to turn water into steam.

The magnetic and inertial confinement schemes are the two most promising approaches to controlled thermonuclear fusion. Although the detailed technology for each approach is somewhat different, the features of fusion power production from either approach look attractive. There are sufficient resources of fusion fuels for either process to satisfy man's energy needs for billions of years. Furthermore there is some reason to hope that nuclear fusion will exhibit significant environmental and safety advantages over alternative energy sources such as fossil fuels or nuclear fission. In fact, it has been suggested that nuclear fusion may offer the best and perhaps the only long-term solution to the complex set of energy-related problems that man will face in the future.[11] The enthusiasm for fusion power has spread from scientists to the public and to those involved in energy policy. On rare occasions one even hears the suggestion (although not from fusion scientists) that we divert resources from our national effort to develop nuclear fission power (particularly the fast breeder reactor) and coal technology to aim a significantly larger effort toward developing nuclear fusion power.[12]

But nuclear fusion is not yet a scientifically feasible energy source, much less a technologically viable source. The fundamental scientific experiments that will demonstrate that a controlled thermonuclear fusion reaction can generate more energy than it consumes during its production have yet to be performed. Nuclear fusion is in the unique position of being the only technology that has been identified as an energy option long before it has been demonstrated that it can result in net energy production.

Research activities with peaceful applications of nuclear fusion were begun in parallel with the thermonuclear weapons program in this country as well as in the USSR and the United Kingdom in the early 1950s.[13] This effort was declassified by international agreement in 1958, and since that time there has been extensive cooperation among the nations in fusion research. Although there was initially a very high degree of optimism that a successful nuclear fusion reactor could be developed at an early date, as actual fusion experiments began to be performed, it became evident that the high-temperature form of the fusion fuel was a considerably more mysterious substance than scientists had originally thought. The plasma fuel was extremely difficult to confine in any device aimed at generating fusion power. Although a significant amount of fusion research effort continued, prospects for an early solution dimmed during the early 1960s.

The announcement in 1966 by the Soviet Union of significant advances using a new type of magnetic confinement device, the Tokamak, gave new impetus to the fusion effort during the late 1960s. Subsequent experiments using this type of machine were performed in other countries including the United States. The declassification of the laser fusion or inertial confinement effort during the early 1970s accelerated the fusion program. The support for fusion research in this country has grown to a level of roughly $500 million a year. This program stands on the threshold of demonstrating the scientific feasibility of the fusion approach to energy production. It has captured the attention and the enthusiasm of a large segment of the scientific community.

Basic Concepts of Fusion

To understand the various approaches to fusion and their likelihood of success, let us consider the underlying physics of thermonuclear fusion reactions.[14] To begin with, only fusion reactions among the lighter nuclei are of interest for practical applications because of the strong electric repulsion between charged particles. To overcome this repulsion and approach each other closely enough for fusion, nuclei must collide at high velocities. The larger the nuclear charge, the more energetic must be the collision. Since the lighter nuclei have the smallest charge, it is easiest to get these to react. But even the lightest nuclei, the isotopes of hydrogen, must collide with each other with enormous kinetic energies for there to be an appreciable probability of a fusion reaction.

One of the most attractive fusion reactions involves the combination of two deuterium nuclei or deuterons to produce helium. Deuterium is present in seawater at a concentration of about 33 grams per cubic meter. Since 1 gram of deuterium has a fusion energy content of 80 billion calories, there is enough deuterium in the oceans to provide man with energy for billions of years. It should be noted, however, that the fusion reaction between deuterium and tritium is far easier to induce than the reaction between two deuterium nuclei (see fig. 22). Unfortunately, the heavier isotope tritium is radioactive with a half-life of 10.3 years and does not occur in nature (aside from a minor contribution due to cosmic radiation incident upon the atmosphere). But tritium can be produced by bombarding lithium with neutrons. In fact, the neutrons produced by the deuterium-tritium fusion reaction itself can be captured by a lithium blanket surrounding the reacting plasma and used to produce tritium directly in the fusion reactor. Therefore the fuel resource base of early fusion reactors will be determined by lithium resources, which are quite large and are not expected to significantly limit the deployment of deuterium-tritium fusion reactors.[15]

The next question is, How we can get these nuclei to run into each other at high enough speeds or energies to induce fusion reactions? This would seem to be a straightforward enough task, since one can easily accelerate charged particles to high energies. For example, the picture tube of your television set accelerates electrons to essentially the required speeds. Therefore we might imagine accelerating tritium ions or tritons to high speeds and bombarding a deuterium target to produce fusion energy. The big flaw in this scheme is that most of the time tritons and deuterons simply bounce off of each other. Such scattering collisions are a million times more probable than fusion reactions. So we are going to have to arrange for tritons and deuterons to collide with each other millions of times or they will probably not fuse together.

One way to bring nuclei up to speeds required for fusion reac-

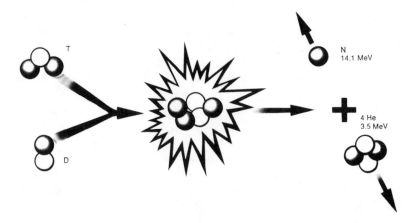

Fig. 22. A deuterium-tritium nuclear fusion reaction. (*Courtesy of the United States Department of Energy.*)

tions is to heat them, for temperature is a measure of the average kinetic energy (and hence the speed) of the atoms in a material. So we must heat the fuel mixture of deuterium and tritium to a temperature such that the average nuclear velocities are large enough to overcome the charge repulsion. But this means that the fuel must be heated to the enormous temperature of 100 million degrees! This is a truly staggering temperature requirement. Certainly no fuel heated to this temperature could remain in a liquid or even a gaseous state. In fact, at these temperatures the atoms ionize, and we have a high-temperature ionized gas that scientists refer to as a plasma. Figure 23 shows the difference between the gas and plasma states. (Although the plasma state of matter may be new to the reader, it is actually the most common state in the universe since it is the substance that constitutes stars. More common terrestrial plasmas include the glowing gases in neon lamps and streetlights.)

Heating the fuel to a sufficiently high temperature to induce fusion reactions, or thermonuclear fusion,[16] is the nuclear analogue of many chemical reactions in which the reacting products must be heated before the reaction will proceed at an appreciable rate.

There is one additional problem. We have seen that the nuclei must collide with each other millions of times before fusion reactions occur. Hence we must contain the fuel at this very high temperature for a period of time if we are to release any appreciable fusion energy. But how do we contain this fuel—more correctly, a plasma—at a temperature of 100 million degrees? The pressures generated by such a plasma are truly enormous. If you were to heat up the air you are breathing to this temperature, it would exert a

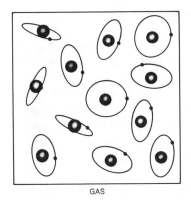
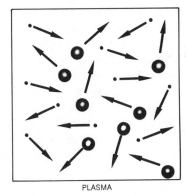

GAS   PLASMA

Fig. 23. Comparison of the gas and plasma states. (*Courtesy of the United States Department of Energy.*)

pressure of 10 million atmospheres. No container on earth could hold it together.

Therefore, to achieve thermonuclear fusion energy release, we must solve two problems: (1) produce and heat a plasma fuel to thermonuclear temperatures, and (2) confine it long enough to produce more fusion energy than we have invested in heating the fuel. These twin requirements are usually quantified by a mathematical relation known as the Lawson criterion,[17] which essentially reflects the balance between thermonuclear energy production and heating energy. This criterion can be expressed as a condition on the product of the fuel density n and the time of plasma fuel containment τ. If we express n in units of number of nuclei per cubic centimeter and τ in seconds, then the Lawson criterion demands that $n \times \tau$ exceed 100 trillion (10^{14}) sec/cm^3 for a deuterium-tritium fusion reaction. The Lawson criterion for a deuterium-deuterium reaction is some 100 times larger (10^{16} sec/cm^3).

This criterion represents only the balance between fusion energy and thermal energy. A Lawson criterion that would characterize a successful fusion reaction, when all of the intrinsic energy losses are taken into account, is some three to five times larger. Therefore the criterion given above is sometimes referred to as scientific break-even, since its achievement will only indicate the scientific feasibility of a thermonuclear fusion scheme, not the engineering viability of the fusion process.

How are we to accomplish the twin goals of heating and confinement in such a way as to satisfy the Lawson criterion? In a star the enormous mass causes gravitational forces that confine the reacting plasma as well as compress and heat it. We cannot expect gravity to do that job here on earth. In thermonuclear weapons no attempt is made to confine the reacting fuel. Rather one merely attempts to heat the fuel to thermonuclear temperatures so fast that an appreciable number of fusion reactions occur before it is blown apart. This scheme is known as inertial confinement since it is the inertia of the reacting fuel that keeps the plasma fuel from blowing apart prematurely. To heat an appreciable mass of fuel to such high temperatures requires an extremely large energy source. The source used in thermonuclear weapons is a fission reaction, that is, an atomic bomb. This approach is highly unsuitable for a controlled application.

The approach to fusion power that has been studied most extensively works with far smaller quantities of thermonuclear fuel. For the past twenty years the primary effort has been to use the charged nature of the plasma fuel as the basis for its confinement in a strong magnetic field. Charged particles have difficulty moving across magnetic field lines and instead spiral along them (see figure 24). Therefore it should be possible to design a magnetic "bottle" in which to contain the plasma fuel.

Magnetic Confinement

Imagine a cleverly designed set of electric coils that generate a magnetic field of such a shape that it can contain a high-temperature plasma. This plasma will exert an outward pressure against the magnetic field, which will in turn cause forces on the electric coils and their support structures. Even for a plasma density as low as one millionth of air at atmospheric pressure, the plasma will exert a pressure of some 10 atmospheres at thermonuclear temperatures. The forces exerted on the magnetic field coils and their supports will be extremely large. Therefore this approach to fusion power will be limited to low fuel density and hence to low power density by the material strength of these components.[18]

Different types and shapes of magnetic bottles for confining thermonuclear plasmas have been proposed and studied. These may be generally classified as either open or closed field geometries. Magnetic confinement arises because charged particles tend to spiral about magnetic field lines and are constrained from moving across these lines. In an open system the magnetic field lines actually leave the system, and therefore particles moving along these lines can escape. To prevent escape, one constricts the magnetic field lines by making the field stronger at the ends of the device. This acts as a

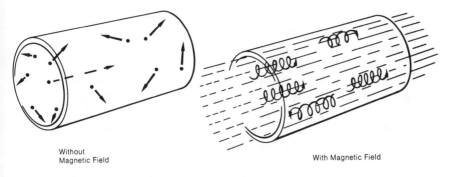

Without
Magnetic Field

With Magnetic Field

Fig. 24. Motion of charged particles in a magnetic field. (*Courtesy of the United States Department of Energy.*)

magnetic "mirror," reflecting particles back into the center of the fields. An alternative approach is to close the magnetic field lines by creating a toroidal geometry, that is, a donut-shaped geometry in which the field lines close upon themselves.

Simple confinement geometries are not sufficient to confine the plasma because the electric and magnetic fields exerted by the motion of the charged plasma particles act to push the plasma out of the magnetic bottle. One fusion scientist has likened the task of confining a plasma in a magnetic field to that of holding a blob of gelatin in a web of rubber bands. The plasma will push against the field lines, distorting them, and eventually will leak out. Therefore more sophisticated field geometries must be used for closed and open magnetic confinement devices.

To overcome the intrinsic leaks from magnetic mirror devices, the original configuration of the simple magnetic bottle was redesigned to achieve a magnetic well geometry in which the magnetic field seen by a charged particle is weakest at the center of the device and strengthens as particles move outward toward the edge. Such configurations are known as minimum-B geometries and can be produced by arranging the magnetic field coils in a geometry similar to the seams on a baseball. A mirror fusion device would operate in a steady-state mode; continuous energy input would be required to sustain the thermonuclear reaction. This could be accomplished by injecting into the device an energetic beam of neutral particles, which would then ionize as they interact with the plasma fuel. The combustion products of the fusion reaction must then be continuously removed. Since a significant amount of the power leaving a mirror device is in the form of energetic charged particles (for the fuel cycles used in these devices), most reactor designs based on this concept employ a direct energy conversion cycle that extracts energy from the escaping charged particles by slowing them down in an electrostatic field (although a thermal energy conversion cycle may also be used with this system).[19]

Two improvements in mirror design have been introduced to reduce particle losses from the ends of the open mirror geometry. One scheme (the field-reversed mirror) involves off-center fuel injection which creates a plasma current to close off open magnetic field lines and trap plasma particles. A second approach (the tandem mirror) involves maintaining small mirror plasmas by strong neutral atom beam injection at either end of a larger mirror configuration to plug end losses. Progress in the heating and confinement of mirror plasmas has been sufficiently encouraging during recent years to stimulate the construction of the large Mirror Fusion Test Facility scheduled for completion in Livermore, California, in 1981. This device should achieve ion temperatures of 500 million degrees and a density-confinement time product within a factor of ten of the Lawson criterion.

A much different approach utilizes a pulsed magnetic field for heating and confining the plasma.[20] In the most common pulsed field schemes, called the theta pinch, an axial magnetic field is induced in the device by discharging a current through a conductor wrapped around the plasma. This field constricts or pinches the plasma, compressing it to higher densities and thermonuclear temperatures. Although a significant amount of work has been done on pinch devices, including the large toroidal Syllac pinch machine at Los Alamos, this particular geometry is still plagued by instabilities that cause the plasma to leak rapidly out of the field. Various "stoppered" linear pinch geometries are now being studied as alternatives to toroidal theta pinch geometries.

The most successful magnetic confinement approach to date utilizes a toroidal geometry in which additional coils are wrapped around the toroid to induce a current in the plasma, producing a shear in the magnetic field lines.[21] This results in an average minimum-B effect that leads to increased confinement times. This particular approach was developed by the Soviet Union and is referred to as the Tokamak (to = toroidal, ka = chamber, and mak = magnetic). Since the successful Soviet experiments with this approach during the late 1960s, a large number of these devices have been built around the world in almost a bandwagon movement.

Progress with the Tokamak approach has been quite impressive. In 1978 a hydrogen plasma in the Princeton Large Torus device at Princeton was heated using neutral deuterium beam injection to the 60 million degree temperatures necessary for a reactor. Although this device was characterized by a relatively low density of 10^{13} cm^{-3}, the Alcator Tokamak device at MIT has attained higher fuel densities at lower temperatures leading to an $n\,\tau$ product of 3×10^{13} sec/cm^3. These results should be compared with n values of ~ 3 to 5×10^{14} sec/cm^3 and temperatures of 60 million degrees needed for a Tokamak power reactor. It seems highly probable that the next large Tokamak experiment, the Tokamak Fusion Test Reactor scheduled for completion at Princeton in 1982, will achieve these goals and demonstrate the scientific feasibility of fusion power by the mid-1980s (fig. 25).

In summary, magnetic fusion research has reached a stage where plasma temperatures and confinement times are within modest factors of those required for break-even. The objectives of this

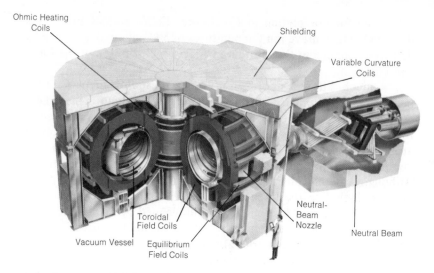

Ohmic Heating Coils

Shielding

Variable Curvature Coils

Toroidal Field Coils

Vacuum Vessel

Equilibrium Field Coils

Neutral-Beam Nozzle

Neutral Beam

Fig. 25. The Tokamak Fusion Test Reactor under construction at Princeton, New Jersey. (*Courtesy of the United States Department of Energy.*)

program are to demonstrate actual reactor-level conditions.[22] However, there are significant aspects of fusion plasma behavior such as confinement times and scaling properties that are still inadequately understood. Much scientific research remains to be conducted before the magnetic confinement fusion program can be redirected toward the engineering development of a power reactor (fig. 26). See table 20 for the present timetable.

Fusion Reactor Concepts
Although there is still a great deal of uncertainty about the specific design of a fusion power reactor, fusion reactors are likely to be large and expensive. The most successful plasma containment geometry, the Tokamak, is characterized by rather low power densities. This implies large plasma volumes. Furthermore such systems will involve an extremely complex and novel technology, even compared with present-day fission reactors.

Let us describe the preliminary conceptual design of a fusion power reactor based on the Tokamak concept.[23] The Tokamak reactor is contained in a thick primary containment shield, which is housed in a secondary containment structure adjacent to a building containing turbines and electric generators. The Tokamak reactor torus is surrounded by a complicated blanket of stainless steel cells containing liquid lithium or solid lithium oxide pellets. This material

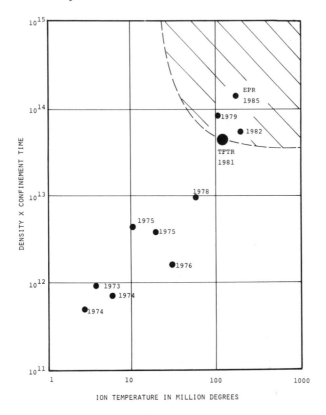

Fig. 26. Progress in fusion research plotted in a graph showing density multiplied by confinement time versus ion temperature. Each dot represents a particular fusion experiment or device. Shaded area represents scientific breakeven.

is designed to capture the energy of the neutrons produced in the deuterium-tritium fusion reactor and also to use these neutrons to produce tritium, which can then be separated out and later reinjected into the plasma fuel. Figure 27 shows a drawing for a proposed experimental fusion power reactor.

Such a reactor would utilize a semisteady-state burn cycle that begins when a large transformer produces an electric discharge in the deuterium-tritium gas contained in the torus to produce a plasma and build up a current in the torus. A number of large neutral beam injection guns are then fired into the plasma, injecting both high-energy deuterium and tritium. These beams inject fuel into the plasma and also raise the plasma fuel temperature to the thermonuclear ignition point. As the thermonuclear fusion reaction begins, the

TABLE 20. Timetable for Controlled Thermonuclear Fusion Development

	1975	1980	1985	1990	1995	2000
Magnetic Confinement Tokamak Mirror		Reactorlike conditions (Princeton Large Torus, Alcator)	Deuterium-tritium reactor feasibility (Tokamak Fusion Test Reactor, Mirror Fusion Test Facility)	Prototype experimental power reactor	Commercial-scale demonstration plant	
Inertial Confinement Laser Fusion Electron Beam, Ion Beam Fusion		High-density implosions (Shiva, Helios, Electron Beam Fusion Accelerator-I)	Feasibility (Nova, Antares, Electron Beam Fusion Accelerator-II)	Prototype power reactor	Commercial-scale demonstration plant	

Fig. 27. The Experimental Power Reactor based on the Tokamak concept which has been proposed for construction at Oak Ridge, Tennessee. (*Courtesy of the United States Department of Energy.*)

alpha particles released in the fusion reactions heat the plasma fuel still further until it reaches the designed operating level where the reactor will continue to operate for some ninety minutes. During this period the reactor is continually refueled by injecting small pellets of deuterium and tritium ice into the plasma. The burn cycle time is limited by the ability of the transformer to sustain the toroidal plasma current, as well as the buildup of impurities in the plasma fuel. At the end of the burn cycle, impurities are injected into the plasma to shut off the fusion reaction. The induced currents and magnetic field are reduced, and the reactor chamber is pumped empty of all gases and then refilled with fresh fuel to ready it for the next cycle.

The neutron energy deposition in the lithium blanket acts as a volumetric heat source that is withdrawn by a secondary coolant such as sodium or helium and used as the heat source for a steam thermal cycle to generate electric power. Some 20 percent of this electric power is used to provide magnetic fields for confinement and the toroidal current and to drive the neutral beam injection. Therefore a considerable amount of energy is circulating within such a nuclear fusion reactor plant.

The successful design of a fusion reactor will require the solution of a number of rather imposing technical problems. Perhaps the most severe problem involves the extensive radiation damage done to the wall of the reactor chamber (the "first-wall problem") as well as to structural materials. Indeed the neutron flux generated by a controlled fusion reactor will be almost ten times larger than that of even a fast breeder reactor. To put this in perspective, over the thirty-year operating lifetime of the plant, every single atom in the first wall would be displaced over five hundred times by fast neutron collisions. This leads to significant loss of ductility and swelling in vessel materials. To design a wall material that can withstand such damage is no easy task. It will probably be necessary to replace periodically the wall liner as well as a significant amount of the blanket structural material. Several fusion reactor designs break up the torus into a number of pie-shaped modules, each of which can be removed and replaced when necessary. Another alternative involves the periodic use of chemical vapor deposition in situ to bake on a new first-wall coating. Needless to say, there is strong incentive to develop materials and first-wall designs that will extend wall life. But there is still no structural material and operating condition that will ensure a vessel lifetime comparable to that of the plant.

The extremely high neutron flux levels also induce a substantial amount of radioactivity in materials adjacent to the reactor chamber.

Therefore maintenance operations will be difficult and will require remote handling techniques. A variety of other materials problems arise, including thermal stress resulting from the frequent temperature pulses in fusion reactors.

A fusion reactor does exhibit an intrinsic safety advantage over fission systems.[24] There can be no runaway of the fusion reaction since it can only burn the amount of fuel that is in the reaction chamber at one time (less than 1 gram). Nevertheless the radioactivity inventory in the plant will be a considerable hazard. The first generation of fusion reactors will be based on the deuterium-tritium fuel cycle that produces an intense flux of fast neutrons. This neutron flux will induce or activate a substantial radioactivity in both the first wall and structural materials. The radioactivity inventory will be comparable to, although somewhat less than, that of a fission reactor. Furthermore this radioactivity will lead to a decay heat removal problem that is qualitatively similar to that of fission reactors and will require emergency cooling similar to the emergency core cooling system on large fission power plants. However, decay heat will be from one to two orders of magnitude below that of the fission products in fission reactors, and therefore emergency cooling should be considerably easier.

The activated structural materials in a fusion reactor will also give rise to a radioactive waste problem, since the half-life of this radioactivity is usually several decades. Activated structural material will retain its toxicity for at least a century. Furthermore a fusion power plant will produce a volume of radioactive waste comparable to that produced by a fission plant. However, in terms of radioactive hazard, this will still be at least an order of magnitude below that produced by fission reactors. The use of more exotic materials such as niobium and vanadium has been proposed for first wall designs which would significantly reduce the neutron activation problems now presented by stainless steel.[25]

Perhaps the most serious hazard associated with fusion reactors involves the substantial inventory of radioactive tritium that these reactors will contain. Although the tritium inventory in the reacting plasma itself will be less than 1 gram, the total tritium inventory in the plant, including the lithium blanket and the tritium recovery system, may be as high as 10 kilograms,[26] although some proposed designs may reduce this inventory by a factor of 10.[27] The escape of even a small portion of this inventory from the plant would constitute a major radiological hazard since tritium is easily assimilated by biological organisms (as tritiated water). Safe operation of the plant will require that the routine release of tritium be kept extremely low.

This is difficult, since tritium diffuses quite readily through most metals at high temperatures and therefore can diffuse through containment walls or fluid piping into the surrounding atmosphere or into the coolant and steam supply systems. To achieve a tritium release rate from fusion power plants comparable to that of present-day fission reactors will necessitate an improvement in tritium handling techniques that will achieve a tritium containment in the blanket and fluid piping in excess of 99.9999 percent. Although the technology is available to do this, the implementation of advanced tritium handling technology will almost certainly have a major impact on the capital cost of fusion plants.

Fusion reactor designs not only involve a remarkable increase in complexity over present fission reactor systems, but they also require a rather significant development of new technology if the problems posed by radioactive materials control, radiation damage, and effective plant maintenance in high-radiation environments are to be solved. One serious drawback of present magnetic fusion reactor designs is the low power density of these systems.[28] For example, the plasma fuel power density in most Tokamak designs ranges from 1 to 10 megawatts per cubic meter as compared to a core power density of 100 megawatts per cubic meter for light water reactors and 400 megawatts per cubic meter for fast breeder reactors. The serious implications of this lower power density become apparent when it is recognized that fusion energy produced by the reacting plasma must be collected outside of the plasma region, not extracted by flowing a coolant through the reaction volume as one is able to do with fossil-fuel boilers or fission reactors. All energy must be gathered outside the reacting plasma region, and the required area of surrounding surface must have associated with it a thick structure of complex design. Therefore the periphery of the plasma volume determines the size of the heat extraction or blanket region. Since capital cost is closely related to the physical size of the nuclear steam supply system, this low power density may place fusion reactors at a decided economic disadvantage.

There is a strong incentive to develop reactor designs with higher power densities, but increasing power densities will produce new difficulties. It is more difficult to confine higher power density plasmas. Furthermore the radiation fluence passing through the reactor vessel walls becomes more intense and therefore causes more damage to vessel components. It is more difficult to remove the heat from the periphery of reacting plasmas at higher power densities.

By employing plasmas of noncircular cross-section or so-called flux-conserving Tokamak designs,[29] one may be able to increase power density. It is certainly premature to conclude that fusion reactors have to be huge. However, early fusion reactors probably will operate at only about one-third the power density of a breeder reactor.[30] Furthermore fusion reactors tend to require more scarce and expensive materials. These factors suggest that the capital cost of the nuclear steam supply system for a fusion plant will be significantly higher than that for a fission plant.[31] Availability of exotic materials such as niobium, vanadium, and chromium may constitute a primary resource base limitation to the rapid expansion of fusion power should it become a viable technology.

Although there was early optimism that fusion power might exhibit significant economic advantages over fossil fuels, nuclear fission, and renewable sources such as solar power, these hopes have faded as the technological complexity of fusion reactors has become more apparent. The inherently low power density in fusion systems and the possible necessity of replacing key components of the fusion reactor structure at regular intervals due to radiation damage make it very unlikely that the nuclear island of a fusion plant will be comparable in cost to that of a fission plant. It is now estimated that fusion reactors will cost perhaps several times more than fast breeder reactors, leading to total plant capital costs in the range of a few thousand dollars per kilowatt capacity.[32] It is almost certain that if fusion power is chosen as a significant component of our future energy supply, it will not be for an entirely economic reason, but rather because of its superior environmental or safety features. It has become apparent over the past twenty years of research that the successful development of controlled fusion power, if it occurs, will stand as one of the major scientific and technological accomplishments of our civilization.[33] The difficulties of achieving practical fusion power cannot be understated. For this reason there has been recent interest in combining nuclear fusion with conventional fission reactor technology to reduce the design requirements of a fusion system.[34]

One distinguishing characteristic of fusion reactors is that they will be neutron rich, that is, they will produce roughly four times as many neutrons per unit energy output as a fission reactor. Hence considerable thought has been given to using the fusion reaction not so much as a power source, but rather as a high-intensity neutron source, which can then drive a subcritical blanket of fissile material surrounding the fusion reactor. Fusion energy production can be multiplied by the fission energy induced by fusion neutrons. Furthermore these fast neutrons can breed new fissile material from fertile material, so we might visualize a fusion system as a neutron source producing fuel for conventional fission reactors. Such hybrid fission-fusion systems certainly relax somewhat the requirements for achieving a break-even fusion reaction. However, they also may combine the bad points of both fission and fusion systems, such as high inventories of radioactive materials, plutonium, the complexity of achieving a sustained thermonuclear fusion reaction, high tritium inventory, and so on.

The use of fusion reactors as intense neutron sources may have other applications. The fusion neutrons could be used to convert fertile material such as uranium-238 or thorium-232 into fissile material such as plutonium or uranium-233 for use in conventional fission reactors. Of course, this application would encounter the same concern for the international proliferation of nuclear weapons capability that has been stimulated by proposals to implement the fast breeder reactor. Fusion reactions could also be used to transmute long-lived radioactive waste (actinides) into shorter lived or stable isotopes.[35] Yet another application would be to use the neutrons produced in a fusion reactor to produce chemical fuels, say by radiolytically decomposing water into hydrogen and oxygen and then using the hydrogen in chemical processes to produce methane that could supplement our vanishing natural gas reserves.

Inertial Confinement Fusion

An alternative approach to achieving controlled thermonuclear fusion involves heating a tiny pellet of fuel (for example, a frozen droplet of deuterium and tritium about the size of the head of a pin) to thermonuclear temperatures so rapidly that it ignites and burns through thermonuclear reactions, releasing fusion energy before it can blow itself apart. In this scheme the only confinement of the burning plasma fuel is provided by its own inertia. Here the premium is placed on the rapid heating of the pellet, which is accomplished by zapping the pellet with high-intensity laser beams or charged particle beams.

Inertial confinement fusion can be regarded as essentially the internal combustion approach to fusion.[36] To make the analogy more precise, recall that the internal combustion engine of your car is based on a four-stage combustion cycle: (1) injection of fuel (gas and air) into the cylinder, (2) compression of the fuel mixture by a piston, (3) ignition of the compressed fuel by a spark plug, and (4) combustion of the fuel mixture in a small explosion that drives the piston and hence the crankshaft (conversion of chemical energy to mechanical energy).

In direct analogy, inertial confinement fusion schemes are based on the following sequence: (1) a tiny pellet of deuterium-tritium isotopes is injected into a blast chamber, (2) the pellet is compressed to very high density with intense laser, electron, or ion beams, (3) the high density and compression heat induces the ignition of a thermonuclear reaction, (4) the thermonuclear energy carried by reaction products including neutrons, X rays, and charged particles is deposited as heat in a blanket that then acts as a heat source in a steam thermal cycle that produces electricity (conversion of nuclear energy into electric energy). Figure 28 compares a fusion reactor with the internal combustion engine. The laser fusion internal combustion engine will use a series of microthermonuclear explosions (from 10 to 100 per second, each generating the energy equivalent of several pounds of high explosive) to generate electric power.

The general ideas behind inertial confinement schemes have been around for a long time, at least since the discovery of the laser in 1960. However, because of their intimate relationship to the physics of nuclear weapons, most relevant details of the scheme were shrouded in a blanket of security classification until roughly 1972. Largely through the efforts of scientists in the nuclear weapons program, many details of laser fusion have been de-

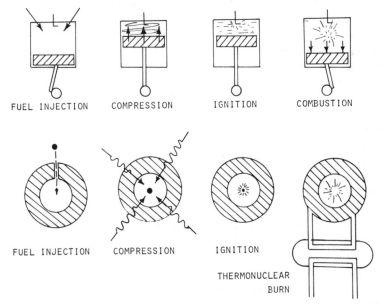

Fig. 28. A comparison of internal combustion engines: the top sequence for a gasoline engine of an automobile, the bottom sequence for a laser fusion reactor.

classified, and more recently a considerable amount of research information on this approach to fusion has shifted into the open literature.[37]

To explain how this approach to fusion works, let us first recall that in magnetic confinement approaches one attempts to beat the Lawson criterion for fusion feasibility, that is, $n \tau > 10^{14}$, by confining a low density plasma with $n \sim 10^{14}$ for a relatively long time, $\tau \sim$ 1 second. The inertial confinement fusion scheme takes essentially the opposite approach. Here we attempt to heat a dense fuel to thermonuclear temperatures extremely rapidly so that an appreciable thermonuclear reaction energy will be generated before the fuel blows itself apart. To see what we are up against, consider a small pellet with a radius of 1 mm. The disassembly time τ required for the heated pellet to blow itself apart is roughly just the time required for a sound wave to traverse the pellet. Since the speed of sound in a thermonuclear plasma is roughly 10^8 cm/sec, the disassembly time is $\tau \sim 0.1/10^8 = 10^{-9}$ sec or 1 nanosecond. Hence, to satisfy the Lawson criterion, we must use a fuel density in excess of $n = 10^{14}/\tau = 10^{23}$ cm^{-3}, which is roughly the density of a solid.

The new game we must play in inertial confinement is to heat a small, high-density fuel pellet to thermonuclear temperatures before it has a chance to expand, that is, in 1 nanosecond or one billionth of a second! But how can we heat the fuel this rapidly? This is where the laser comes in. For not only can a laser focus large amounts of energy on tiny spots, but it can also zap this energy in a very short time, easily within one nanosecond. Laser pulses as short as one trillionth of a second have been achieved.

So if we use the laser as a big flashlight to zap the fuel pellets to fusion temperatures rapidly, we can induce a thermonuclear microexplosion. The energy from this explosion can then be captured and converted into electricity through a steam thermal cycle. Part of this energy is used to reenergize the laser, and the rest is distributed to the electric power grid. Figure 29 is a schematic of laser fusion reactor.

So far, so good! And this was essentially the public image projected by the laser fusion effort in the BDC (before declassification) days prior to 1972.[38] But this simple-minded scheme had a fatal flaw that became apparent when one tried to estimate the laser energy required to produce such a microexplosion.

Performing a simple energy balance, one finds that the laser energy required to ignite a solid pellet of deuterium-tritium fuel so that it produces an equivalent amount of fusion energy (scientific feasibility) would be almost 10 million joules. A similar estimate for a

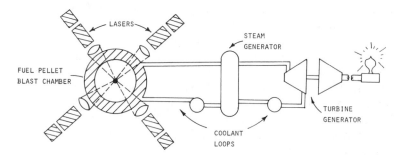

Fig. 29.   A simple schematic of a laser fusion reactor.

reactor (technical viability) increases this to 10 billion joules. To place these numbers in perspective, we should note that the largest laser in the world today, a $20 million monster known as Shiva at the Lawrence Livermore Laboratory in Livermore, California, can produce a pulse of only 20 thousand joules, a thousand times too small. So viewed in this light, laser fusion is clearly a fool's quest.

Or is it? This was the naive or BDC approach. We must be a bit more sophisticated in our analysis.[39] Let us reexamine the criterion for achieving net fusion energy release in a somewhat different light. Two times are of major significance for inertial confinement schemes: the disassembly time required for the pellet to blow apart and the thermonuclear burn time required for an appreciable number of thermonuclear reactions to occur. The efficiency with which the pellet will burn is clearly related to the ratio of these times. This efficiency can be estimated in terms of the pellet density ρ in grams per cubic centimeter and the pellet radius R in centimeters as $\rho R/(6 + \rho R)$. To understand the implications of this result, note that for a 0.1 centimeter pellet, $\rho R = 1$ implies a fuel density of $\rho = 10$ g/cm^3. But since solid-state density of a deuterium-tritium mixture is only $\rho_s = 0.2$ g/cm^3, we must somehow *compress* the pellet to at least fifty times its solid density.

The key to inertial confinement fusion is apparently high compression. The more we compress the fuel, the larger ρR becomes, and hence the more efficient the thermonuclear burn and the larger the energy yield. The required laser energy for break-even decreases as the square of the compression factor increases. To see how this affects our earlier estimates of required laser energy, if we can achieve a compression of one thousand times solid density, we find that the break-even energy required is only 10 joules, while the reactor energy is now 100 joules. Actually these estimates are a bit too optimistic, since they ignore the fact that a good deal of laser light incident on the pellet will simply be reflected off. Nevertheless they do indicate the required laser energy is inversely proportional to the square of the pellet fuel compression.

The only remaining question is, How do we achieve such enormous compressions? Certainly not by normal mechanical forces. Nor will chemical explosives do the job, since they are limited to compressions of about 10 by the strength of interatomic forces. Densities as large as one thousand times solid density are not common even on an astronomical scale and occur only in dense white dwarf stars.

The trick involves using the laser itself. The basic scenario goes as follows: The intense laser light is focused by a number of laser beams onto the pellet surface. As the pellet absorbs this intense light, its surface rapidly vaporizes, ionizes, and heats to high temperature, blowing off into the vacuum surrounding the pellet. This blowoff or ablation of the pellet surface drives a shock wave back into the pellet (recall Newton's third law—or better yet, picture the ablation as you would the thrust from a rocket). As this shock wave implodes toward the center of the pellet, it compresses the fuel to high density and thermonuclear temperatures so that ignition of the thermonuclear burn occurs. At these very high densities, the energetic alpha particles produced in deuterium-tritium fusion reactions are absorbed in the fuel, heating it to still higher temperatures and causing the fuel to burn even more rapidly. After only a few picoseconds (10^{-12} seconds), a significant fraction of the imploded fuel pellet has burned, and the high energy release blows the pellet apart, thereby terminating the reaction.

Although this scheme sounds farfetched, it has been demonstrated in laboratory experiments. In experiments, laser beams are focused by specially designed mirrors onto the surface of pellets that consist of tiny glass shells (50 microns in diameter and 1 micron in thickness) containing deuterium-tritium gas at up to 10 atmospheres pressure (see fig. 30). By carefully studying the X rays emerging from these irradiated pellets, scientists have verified compressions of the initial fill gas density as large as 1,000 (about 100 times solid density), accompanied by the emission of thermonuclear neutrons.[40]

The success of such experiments should not be interpreted as a demonstration of the scientific feasibility of laser fusion. Present estimates are that the achievement of the $\rho R > 1$ g/cm^2 required for efficient thermonuclear burn will require an absorbed laser energy of roughly 1,000 joules in the pellet delivered in such a manner as to induce a compression of some ten thousand times solid density. For a laser fusion reactor, the requirements become even

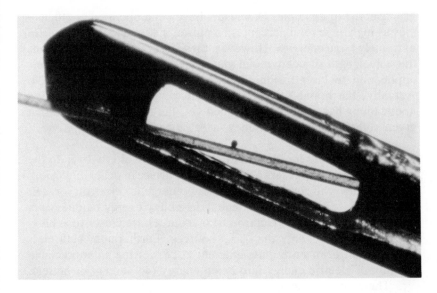

Fig. 30.   A tiny laser fusion fuel pellet suspended on a human hair in the eye of a needle. (*Courtesy of the United States Department of Energy.*)

more severe, with $\rho R > 5$ g/cm² corresponding to an absorbed laser energy of tens of thousands of joules. Furthermore many aspects of the coupling of laser light energy to hot plasmas are still not adequately understood. It is difficult to maintain an adequate spherical symmetry under these strong compressions. Much of the incident light is reflected off the surface of the pellet, and the fabrication and positioning of these tiny pellets are quite complex.

The energies required for laser fusion present a difficult challenge to high-powered laser technology. Although 10,000 joules may not sound like a great deal of energy (it is about the release of a small firecracker), when it is delivered in 10^{-10} seconds it corresponds to a power level of 10^{14} watts—roughly one hundred times the instantaneous electric generating capacity of the United States. Most high-energy laser facilities designed for laser fusion research utilize large, neodymium glass lasers that emit infrared light. To date these lasers have been restricted by glass damage to energies less than 1,000 joules per beam. Several laboratories in the United States and abroad have large multibeam neodymium glass lasers in the 10,000 joules per pulse range.

These lasers are rather inefficient (0.1 percent) since they must be energized with flash lamps. To achieve the high efficiencies required by reactor applications, it will probably be necessary to use gas lasers. For example, CO_2 lasers have been operated at efficiencies as high as several percent (in a pulsed mode) and are capable of extremely large energies. However, these lasers operate at ten times the wavelength of neodymium glass lasers, which may make their application to laser fusion somewhat more difficult (see fig. 31). Actually, the brand-X laser that is most suitable for laser fusion applications has not been developed yet, but the assumption is that given sufficient time and money laser physicists can design one.[41]

Electron Beam and Ion Beam Fusion.   There are alternatives to the use of high-powered lasers as the piston or driver in pellet implosion. Accelerators can produce extremely energetic electron or ion beams.[42] If we compare the scientific feasibility energy requirements with present capability, electron beam sources are closer to achieving the required energy than laser sources. Furthermore both electron and ion beams are quite efficient at converting an appreciable fraction of electric energy into beam energy (40 percent or better).

There are problems however. For example, our usual understanding of electron interactions would predict that high-energy electrons would penetrate right through a pellet without depositing energy to drive the implosion. Fortunately experiments have indicated that these electrons are readily absorbed by the pellet for some

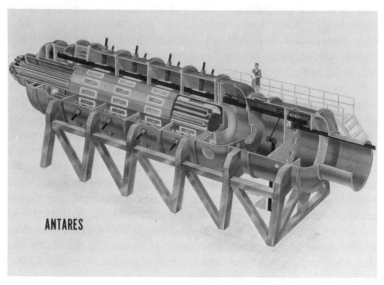

ANTARES

Fig. 31.   A schematic of one module of the Antares laser system under development at the Los Alamos Scientific Laboratory. (*Courtesy of the United States Department of Energy.*)

reason we do not fully understand. This anomalous behavior affects the design of suitable pellets. Furthermore the development of electron beams with the short pulse lengths or high repetition rates required for pellet implosion is not easy.

To overcome the absorption problem one can go to heavy charged particles such as ions. Preliminary studies considered using proton or deuteron beams to compress the pellet. Once again these beams penetrate to the core of the pellet too readily, thereby preheating it and reducing the compression. More recent proposals have been based on using beams of heavy ions such as iodine or even uranium. One could use storage rings in which groups of energetic ions can be accelerated and stored until they are diverted out and focused onto the pellet target. Although the status of heavy ion beam sources is still far from the requirements of pellet fusion, such a scheme might hold some potential for inertial confinement fusion.

Reactor Concepts. Let us leave the question of just how such thermonuclear microexplosions can be generated and consider how such explosions can be used to produce useful energy in a reactor device. Typically pellet implosion is assumed to yield some 10^8 joules (about 50 pounds of high explosive equivalent worth of energy). If such explosions are repeated thirty times a second, the reactor will yield 3,000 megawatts of thermal power corresponding to a thermal cycle output of 1,000 megawatts of electric power.[43]

Thermonuclear explosion energy appears as various types of radiation emitted from the exploding pellet. Typically for a deuterium-tritium pellet 75 percent of the energy will appear as fast neutrons, 24 percent as energetic charged particles, and 1 percent as X rays. Surprisingly enough, it is relatively easy to design a chamber that can withstand the force of such a blast (fig. 32). This is because the force generated on the walls of the chamber is proportional to the square root of the explosion debris mass, and since a thermonuclear explosion utilizes a mass almost a million times smaller than a chemical explosion of similar energy yield, the blast force is rather small (a firecracker's worth).

The principal concern is the damage that the incident radiation can do to the chamber wall. For example, soft X rays and charged particles can damage the wall surface, spalling it off into the blast chamber. The energetic neutrons will cause significant damage to both wall and structural materials. By careful design, such as wetting the wall surface with a thin film of lithium to absorb the X rays and charged particles or shielding the walls from charged particles

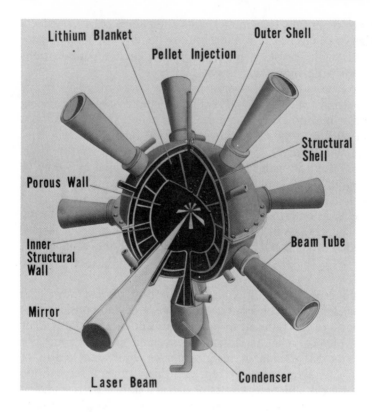

Fig. 32. A laser fusion reactor blast chamber. (*Courtesy of the United States Department of Energy.*)

with magnetic fields, a blast chamber should be able to contain such pellet implosions.

Most explosion energy would be carried by fast neutrons; therefore the blast chamber would be surrounded by a blanket, such as lithium, to absorb the neutron energy (and perhaps produce tritium for further refueling). This blanket would then be cooled by conventional techniques, and the heat withdrawn by the coolant would be used to produce steam for a turbogenerator. Figure 33 is an artist's drawing of a laser fusion power plant.

The successful development of inertial confinement fusion reactors faces many of the same problems as the magnetic fusion approach. Radiation damage to reactor components will be a significant problem. Radiological concerns involving tritium handling or neutron activation are comparable. The design of efficient and reliable laser or charged-particle accelerator drivers will pose a formid-

Fig. 33. An artist's conception of a laser fusion power plant. (*Courtesy of the United States Department of Energy.*)

able problem. And the security classification of high-gain pellet designs poses a significant barrier to wide-scale commercialization of this approach.

Whatever the design, the successful development of a viable laser (or electron or ion beam) fusion reactor is still many years down the road. Indeed, we are still several years away from demonstrating the scientific feasibility of inertial confinement, just as we are with magnetic confinement.

The Future of Fusion Power

The early history of nuclear fusion research was buoyed by the expectation that this process would provide a virtually inexhaustible supply of clean, cheap energy. Nuclear fusion reactors could use as fuel the almost unlimited quantities of deuterium in the oceans. Since these reactors would contain only a small quantity of fuel during operation, there would be no danger of a nuclear runaway accident, and therefore a nuclear fusion reactor should be inherently safer than a fission reactor. The total inventory of radioactive material in such reactors would be significantly smaller than that in fission reactors. Since fusion reactors would not utilize materials that could, in and of themselves, be made into nuclear weapons, there would not be the danger of proliferation or diversion of strategic nuclear materials. Fusion systems were expected to improve thermal efficiencies through higher temperature operation and possibly convert fusion energy to electric energy directly.

But bringing nuclear fusion to a stage of scientific feasibility has turned out to be far more difficult than anyone suspected. As the difficulties of achieving controlled thermonuclear fusion have become more apparent, the early hopes for this technology have been replaced by more realistic and pragmatic expectations.[44]

The fuel resource base of early fusion reactors will not be based on deuterium, but rather on lithium reserves, since this element will be converted into tritium to fuel the deuterium-tritium fusion reaction. Present estimates are that domestic lithium reserves are sufficient to fuel fusion reactors for thousands of years, certainly until the technically more demanding deuterium-deuterium fusion reactors can be developed. But nuclear fusion is not alone in possessing an inexhaustible fuel supply. Both solar power and the fast breeder reactor are characterized by essentially infinite energy resources. Probably the primary limiting factor on fusion power development will not be the fuel resources, but rather the resources of exotic materials such as vanadium or niobium, which are required for fusion systems.

What about the safety of fusion systems? The problems of fusion power will be similar to those of fission power. Early fusion reactors will produce large quantities of neutrons that will activate structural and blanket materials adjacent to the reactor and lead to a significant radioactivity inventory. Although the magnitude and character (half-lives and hazard potential) of this induced radioactivity will be somewhat less than that characterizing fission systems, it will nevertheless require serious attention. Furthermore the significant tritium inventory in a fusion system represents a radiological hazard during normal plant operation, comparable to the hazards presented by nuclear fission reactors. Fusion power plants will utilize significant amounts of electric energy just to sustain the fusion reaction, and this large energy circulation within the system could present a safety problem. The use of liquid metals such as lithium poses serious fire and explosion hazards, as it does in liquid-metal-cooled fast breeder reactors. Although the radiological hazards of fusion systems may be somewhat less than those of fission systems, they are nevertheless of sufficient magnitude to require a comparable level of attention to the design of engineered safety systems.

What about environmental impact? Like fission systems, fusion power plants will not release materials such as combustion products to the environment. The release of radioactivity during normal operation also can be limited to the low levels characterizing fission plants. The early generation fusion power plants will utilize a thermal cycle similar to that of fission reactors. Therefore they will be limited to the same thermodynamic efficiencies and waste heat discharges. If the materials problems caused by radiation damage to the first wall of the reactor chamber cannot be alleviated through design, it may be necessary to operate fusion power plants at significantly lower temperatures than advanced fission systems such as the high-temperature gas-cooled reactor or the liquid-metal-cooled breeder reactor. The primary environmental advantages of fusion power involve its fuel cycle, since it does not require the elaborate processing and handling that characterizes fission reactor fuels. Most fuel handling and processing will occur in the power plant itself.

Even though fusion power may not be completely clean and totally safe, it may prove more socially acceptable than fossil fuels or nuclear fission power because of its potential for reduced environmental impact and hazards. (Solar power will certainly be its strong competitor in this regard.) Fusion power also has a decided advantage from the standpoint of nuclear weapons proliferation. Fusion systems do not involve nuclear materials such as plutonium, which could be used directly for weapons fabrication (although since the fusion reaction is a copious neutron source, it could conceivably be used to produce this material). This latter aspect should be qualified with respect to laser fusion since this technology does involve classified features that are related to thermonuclear weapons.

A nuclear fusion reactor will be an extremely complicated and expensive device. The increase in complexity in passing from the present generation of fission reactors to fusion systems is probably comparable to that involved in the transition from coal-fired boilers to nuclear reactors. The engineering problems in making fusion power a viable source of energy are difficult indeed.

Fusion reactors will be at least as expensive as advanced fission reactor types such as the fast breeder, and many studies indicate that the nuclear components of fusion plants will probably be significantly more expensive than conventional fission systems. The primary factors in costs are the low power densities of present fusion designs, the use of sophisticated technologies and materials in these designs, and the necessity for frequent replacement of components damaged by the high-radiation fields in fusion reactors.

We should recognize that while fusion power may present both environmental and safety advantages over alternative energy sources, these will be attained only at a considerable expense and only after significantly more scientific research and engineering development. The fusion program in this country and abroad is now entering that stage when the investment required for further progress will become quite large. For example, the next series of fusion test systems (the Tokamak Fusion Test Reactor and Mirror Fusion Test Facility magnetic confinement reactors and the Nova and Antares laser systems) will cost about $100 to $200 million each. Because of this, there will be the inevitable pressure to narrow the fusion program to a concentration on those concepts that seem to promise the earliest possible commercialization. This trend is being opposed by many scientists[45] who feel that it is still too risky to select a front-running approach such as Tokamak and push it through to commercial viability as rapidly as possible. The fusion program should not proceed to the next development phase of any fusion scheme until experimental results on existing devices warrant it. Large commitments to planned facilities should be based on formal reviews of experimental results obtained with existing facilities. Throughout this development there should be a major emphasis on engineering problems that will provide the basis for proceeding to engineering development and the subsequent choice of an engineering prototype reactor. The fusion approach that offers the shortest route to commercialization may not necessarily be the best choice for our society.[46]

Certainly the development of fusion power is a goal that should be pursued most vigorously. But we should keep in mind that there is little likelihood that fusion power will contribute significantly to the generation of electric power until well into the twenty-first century, probably about 2030. We should not let the glamour of fusion power blind us to its very real difficulties, or prevent us from adequately supporting research into less exotic energy alternatives such as synthetic fuels production, solar power, or the fast breeder reactor.[47]

The Future of Nuclear Power

So what is the future of nuclear power? What role should it play in meeting the energy needs of our society, in helping to slow our rush toward the exhaustion of conventional energy sources? Nuclear power is a necessary component in our future if we are to meet our energy needs and those of developing nations in the face of declining reserves of conventional fossil fuels. There appears to be no viable alternative. Coal production and conservation measures alone cannot be expected to fill the gap created by declining petroleum and natural gas reserves. New energy sources such as solar electric power or nuclear fusion simply will not be available for massive implemention until well after the turn of the century.

A considerable degree of caution and conservatism must be exercised in speculating about the future of nuclear power, however. The massive implementation of this energy source poses enormous difficulties, not the least of which is an adequate public acceptance of this still rather largely misunderstood technology. Equally serious are the enormous capital requirements of nuclear plants and the bewildering complexity of regulations that threaten to delay and choke off further nuclear plant construction. The climate of uncertainty created by confusing government policy changes and inaction on critical issues such as radioactive waste disposal and spent fuel reprocessing have inhibited expansion of nuclear power generation. Furthermore the rather tenuous connection between nuclear power development and nuclear weapons proliferation has given rise to a bitter international debate and stimulated a variety of political attempts to restrict the transfer of nuclear technology, equipment, and materials.

Even if these difficulties can be overcome rapidly, the present generation of nuclear power reactors represents a relatively short-term source of energy, roughly comparable to our fluid fossil fuel resources. Without the introduction of advanced converter or breeder reactors, we will rapidly exhaust our uranium and thorium reserves, and nuclear fission power will cease to be a viable technology shortly after the turn of the century.

If nuclear power is to be more than a temporary energy source, we must look beyond the present generation of nuclear fission reactors to advanced nuclear technologies such as the breeder reactor or controlled thermonuclear fusion. Either of these technologies could supply all of mankind's energy requirements for thousands of years to come. Either can be regarded as an infinite energy source, much as solar power.

The difficulties facing nuclear fusion involve the rather imposing technical tasks of demonstrating the scientific viability of fusion energy production, which is likely to be accomplished during the next decade, and then engineering this highly complex technology into a viable energy source, which is not likely to be accomplished until well after the turn of the century. In sharp contrast, the fast breeder reactor is not only scientifically feasible today, but also technologically viable. Demonstration fast breeder reactor plants have been operating successfully in Europe for several years. Commercial-scale prototype breeder reactors are under construction, and the first commercial sales of fast breeder reactors are anticipated within the next decade.

But a massive implementation of the breeder reactor faces enormous barriers of a nontechnical nature. Foremost among these are the political responses to the fears that breeder reactor technology may accelerate the proliferation of nuclear weapons. This has already led the United States to renounce fast breeder technology and its plutonium fuel cycle in favor of alternative nuclear technologies to set an example for the rest of the world (perhaps an example of its naiveté in believing that a technical fix can be an effective barrier to nuclear weapons proliferation). Then, too, there is an enormous psychological opposition to breeder reactor technology. The potential of these factors for blocking breeder reactor implementation should not be underestimated.

Nuclear power is a highly controversial subject, and opinions concerning its future role differ greatly. It is our belief that nuclear power is essential if our society is to bring into balance its very real needs for energy with its sources of energy supply. But we also acknowledge that there are genuine problems and concerns about nuclear power that must be addressed if this technology is to play a significant role in our future. As nuclear engineers we feel that these problems can be solved, and that nuclear power can be made to be a safe and suitable source of energy.

But there seems little doubt that nuclear power will remain costly. It will continue to concentrate a sophisticated technology in the hands of a few. Certainly, too, a nuclear power station will have a significant impact on its natural environment. And events such as the Three Mile Island accident suggest that nuclear power will never be absolutely safe—nothing is absolutely safe. It will always present some risk to our society. But then the same can be said for all of our future energy options, whether they be nuclear power, or petroleum, or coal, or even solar power. One cannot judge nuclear power in a vacuum; one must instead compare its advantages and disadvantages against those of alternative options.

Perhaps we should be optimistic about the decision-making process in our society and assume that it will carefully weigh the pros and cons of nuclear power, balancing them against the pros and cons of its alternatives, before choosing to accept or discard this technology. We rather suspect that such decisions are more apt to be made under less sensible circumstances. We suspect that if nuclear power is to be accepted and implemented, it will be only as a last resort, when the public realizes that all other options, whether they be coal, oil, or conservation, solar or geothermal power, have been exhausted or recognized as unsuitable or insufficient.

Certainly nuclear power presents us with a trade-off, a balance between benefits and risks. But that is a trade-off that has always been a part of the progress of man.

Notes

Chapter 1

1. R. G. Hewlett and O. E. Anderson, Jr., *The New World, 1939/46* (University Park, Pa.: Pennsylvania State University Press, 1962).
2. Lewis Strauss, *Men and Decisions* (Garden City, N.J.: Doubleday, 1962).
3. *Update: Nuclear Power Program Information and Data, June 1978*, Division of Nuclear Power Development, U.S. Department of Energy (Washington, D.C., 1978).
4. Harrison Brown, "Energy in Our Future," *Annual Review of Energy* 1:1 (1976).
5. M. King Hubbert, "The Energy Resources of the Earth," *Sci. Am.* 224:60 (September 1971); A. R. Flower, "World Oil Production," *Sci. Am.* 238:42 (March 1978).
6. P. H. Abelson, "Public Opinion and Energy Use," *Science* 197:4 (1977).
7. Diana E. Sander, "The Price of Energy," *Annual Review of Energy* 1:391 (1976).
8. *National Energy Outlook*, Federal Energy Administration, (Washington, D.C., February 1976).
9. Earl T. Hayes, "Energy Resources Available to the United States, 1985–2000," *Science* 203:233 (1979).
10. C. E. Whittle et al., *Economic and Environmental Implications of a United States Nuclear Moratorium, 1985–2010*, Institute for Energy Analysis Report ORAU/IEA 76-4 (Oak Ridge, Tenn., 1976).
11. "Electricity Growth Estimates," *EPRI Journal*, Electric Power Research Institute (Palo Alto, Calif., June 1978); *Demand 77: The EPRI Energy Consumption Model and Forecasts*, Electric Power Research Institute Report EPRI EA-621-SR (Palo Alto, Calif., 1978).
12. Amory B. Lovins, "Energy Strategy: The Road Not Taken," *Foreign Affairs* 55:65 (October 1976); Amory B. Lovins, *Soft Energy Paths: Toward a Durable Peace* (Cambridge, Mass.: Ballinger, 1977).
13. *Task Force on Energy Report*, National Academy of Sciences, (Washington, D.C.: 1974).
14. *1976–1986 Electrical Energy Use and Demand*, Detroit Edison Company Report (Detroit, Mich., March 1976).
15. *National Electrical Reliability Council Annual Report*, Edison Electric Institute (New York, 1978).
16. See note 5.

17. B. L. Cohen, "Impacts of the Nuclear Energy Industry on Human Health and Safety," *Am. Sci.* 64:550 (1976); *Reactor Safety Study*, United States Nuclear Regulatory Commission Report WASH-1400 (Washington, D.C., 1975); G. G. Eichholz, *Environmental Aspects of Nuclear Power* (Ann Arbor: Ann Arbor Science, 1976).

18. Gaylord Shaw, "The Search for Dangerous Dams—A Program to Head Off Disaster," *Smithsonian*, January 1978, p. 36; P. Avyaswamy, B. Hauss, T. Hseih, A. Moscati, T. E. Hicks, and D. Okrent, *Estimates of the Risks Associated with Dam Failure*, University of California at Los Angeles School of Engineering Report ENG-7423 (Los Angeles, 1974).

19. D. J. Rose, P. W. Walsh, and L. L. Leskovjan, "Nuclear Power: Compared to What?", *Am. Sci.* 64:291 (1976); R. L. Gotchy, *Health Effects Attributable to Coal and Nuclear Fuel Cycle Alternatives*, United States Nuclear Regulatory Commission Report NUREG-0332 (Washington, D.C., 1977); J. P. McBride, R. E. Moore, J. P. Witherspoon, and R. E. Blanco, "Radiological Impact of Airborne Effluents of Coal and Nuclear Plants," *Science* 202:1045 (1978).

20. H. Inhaber, "Is Solar Power More Dangerous Than Nuclear?", *New Scientist*, May 1978, p. 444; H. Inhaber, "Risk with Energy from Conventional and Nonconventional Sources," *Science* 203:718 (1979); H. Inhaber, *Risk of Energy Production*, Canadian Atomic Energy Control Board Report AECD-1119/Rev-2 (Ottawa, Ont., 1978); Aaron Wildavsky, "No Risk is the Highest Risk of All," *Am. Sci.* 67:32(1979).

21. See note 5.

22. D. Rossin and T. A. Rieck, "Nuclear Power Economics, *Science* 201:582 (1978); L. F. C. Reichle, "The Economics of Nuclear Power" (Paper presented to the New York Society of Security Analysts, New York, 1975 and 1976).

23. F. H. Warren, "Hydroelectric Power," in *Conference on Magnitude and Deployment Schedule of Energy Resources*, ed. W. E. Loveland (Corvallis, Oreg.: Oregon State University, 1975), pp. 37–41.

24. E. D. Griffith and A. W. Clarke, "World Coal Production," *Sci. Am.* 240:38 (January 1979).

25. *A National Energy Plan for Research, Development, and Demonstration*, United States Energy Research and Development Agency Report ERDA-48 (Washington, D.C., 1975).

26. Rose, Walsh, and Leskovjan (note 19); Gotchy (note 19).

27. Genevieve Atwood, "The Strip Mining of Western Coal," *Sci. Am.* 233:23 (March 1975).

28. R. A. Kerr, "Carbon Dioxide and Climate: Carbon Budget Still Unbalanced," *Science* 197:1352 (1977).

29. See note 3.

30. See note 17.

31. Lee Schipper, "Raising the Productivity of Energy Utilization," *Annual Review of Energy* 1:455 (1976).

32. Demand and Conservation Panel of CONAES, National Academy of

Sciences, "U.S. Energy Demand: Some Low Energy Futures," *Science* 200:142 (1978).

33. See note 12.
34. A. M. Weinberg, "Reflections on the Energy Wars," *Am. Sci.* 66:153 (1978).
35. M. Maxey, "Energy Policy: Bioethical Problems and Priorities" (Address to the American Nuclear Society, Ann Arbor, Mich., February 1978).
36. B. Rustin, "Small Is Not Beautiful," *Commentary*, October 1977, p. 70; Llewellyn King, "Nuclear Power in Crisis: The New Class Assault," *Energy Daily*, July 14, 1978, p. 5.
37. William D. Metz and Allen L. Hammond, *Solar Energy in America* (Washington, D.C.: American Association for the Advancement of Science, 1978).
38. W. G. Pollard, "Solar Power," *Am. Sci.* 64:424 (1976); W. G. Pollard, "The Long Range Prospects for Solar Derived Fuels," *Am. Sci.* 64:509 (1976).
39. Allen S. Hirshberg, "Public Policy for Solar Heating and Cooling," *Bull. Atom. Sci.*, October 1976, p. 37.
40. See note 38.
41. See note 38.
42. S. Baron, *Solar Energy: Will It Conserve Our Nonrenewable Resources?* (Oradell, N.J.: Burns and Roe, Inc., 1978).
43. See note 20.
44. See note 25.
45. P. Kruger, "Geothermal Energy," *Annual Review of Energy* 1:159 (1976).
46. A. J. Ellis, "Geothermal Systems and Power Development," *Am. Sci.* 63:510 (1975).
47. R. C. Axtmann, "Environmental Impact of a Geothermal Power Plant," *Science* 187:795 (1975).
48. *Report of the Uranium Resources Group Supply and Delivery Panel*, Committee on Nuclear and Alternative Energy Systems (CONAES), National Academy of Sciences (Washington, D.C., 1978).
49. Georges A. Vendryes, "Superphenix: A Full-Scale Breeder Reactor," *Sci. Am.* 236:26 (March 1977).
50. S. M. Keeny et al., *Report of the Nuclear Energy Policy Study Group: Nuclear Power Issues and Choices* (Cambridge, Mass.: Ballinger, 1977).
51. D. Steiner and J. F. Clarke, "The Tokamak: Model-T Fusion Reactor," *Science* 199:1395 (1978).
52. John P. Holdren, "Fusion Energy in Context: Its Fitness for the Long Term," *Science* 200:168 (1978).
53. W. E. Parkins, "Engineering Limitations of Fusion Power Plants," *Science* 199:1403 (1978); W. D. Metz, "Fusion Research I: What Is the Program Buying the Country," *Science* 192:1320 (1976); W. D. Metz, "Fusion Research II: Reactor Studies Identify More Problems," *Sci-*

ence 193:38 (1976); G. L. Kulcinski, G. Kessler, J. Holdren, and W. Hafele, "Energy for the Long Run: Fission or Fusion?", *Am. Sci.* 67:78 (1979).

Chapter 2

1. R. C. Seamans, "Alternative Energy Systems," *Trans. Am. Nucl. Soc.* 24:3 (1976); *Update: Nuclear Power Program Information and Data, June 1978*, Division of Nuclear Power Development, U.S. Department of Energy (Washington, D.C., 1978).
2. A variety of introductions to nuclear power are available: J. R. Lamarsh, *Introduction to Nuclear Engineering* (Reading, Mass.: Addison-Wesley, 1975); R. L. Murray, *Nuclear Energy* (New York: Pergamon, 1975); T. M. Connolly, *Foundations of Nuclear Engineering* (New York: Wiley, 1978); J. J. Duderstadt, *Nuclear Power* (New York: Marcel Dekker, 1979).
3. J. J. Duderstadt and L. J. Hamilton, *Nuclear Reactor Analysis* (New York: Wiley, 1976), p. 69.
4. A. M. Perry and A. M. Weinberg, "Thermal Breeder Reactors," *Ann. Rev. Nucl. Sci.* 22:317 (1972).
5. Lamarsh (note 2); Duderstadt (note 2).
6. H. C. McIntyre, "Natural-Uranium Heavy-Water Reactors," *Sci. Am.* 233:17 (October 1975).
7. H. Stewart, et al., "The High-Temperature Gas-Cooled Reactor," *Adv. Nucl. Sci. Tech.* 4:1 (1968).
8. G. A. Vendryes, "Superphenix: A Full-Scale Breeder Reactor," *Sci. Am.* 236:26 (March 1977).
9. R. G. Hewlett and O. E. Anderson, Jr., *The New World, 1939/1946* (University Park, Pa.: Pennsylvania State University Press, 1962), p. 714.
10. *Atoms for Power: U.S. Policy in Atomic Energy Development* (New York: American Assembly, Columbia University Press, 1957), p. 46.
11. *The Journals of David E. Lilienthal*, vol. 2, *The Atomic Energy Years, 1945–1950* (New York: Harper & Row, 1964).
12. B. Wolfe, "Some Thoughts on New Energy Sources," *Nucl. News* 19:49 (1976).
13. T. Stevenson, "Gloom on the Monangahela," *Sci. Rev.*, January 1977, p. 6.
14. D. Rossin and T. A. Rieck, "Economics of Nuclear Power," *Science* 201:582 (1978).
15. Energy Reorganization Act of 1974, United States Public Law 93–438.
16. Calvert Cliffs Coordinating Committee v. U.S. Atomic Energy Commission, U.S. District Court of Appeals, District of Columbia Circuit Court, 1971.
17. See note 14.
18. Ebasco Corporation Studies for the New York State Public Service Commission, Albany, N.Y., July 1977.

19. "Atomic Industrial Forum Annual Survey on World Nuclear Power Capacity," *AIF INFO* 107:1 (June 1978).
20. Statement to Congress on energy legislation, President J. Carter, March 1977.
21. R. G. Hewlett, *The Atomic Shield* (University Park, Pa.: Pennsylvania State University Press, 1972).
22. F. H. Schmidt and D. Bodansky, *The Fight over Nuclear Power* (San Francisco: Albion, 1976).
23. Craig Hosmer, "The Anatomy of the Nuclear Power Debate," *Electric Perspectives*, May 1976, p. 22; M. Maxey, "Energy Policy: Bioethical Problems and Priorities" (Address to American Nuclear Scoiety, Ann Arbor, Mich., February 1978); Llewellyn King, "Nuclear Power in Crisis: The New Class Assault," *Energy Daily*, July 14, 1978, p. 5.
24. S. McCracken, "The War Against the Atom," *Commentary*, August 1977, p. 33; S. Ebbin and R. Kasper, *Citizen Groups and the Nuclear Power Controversy* (Cambridge, Mass.: M.I.T. Press, 1974); King (note 23).
25. Sheldon Novick, *The Careless Atom* (Boston: Houghton Mifflin, 1969); J. Gofman and A. R. Tamplin, *Poisoned Power* (Emmaus, Pa.: Rodale Press, 1971).
26. D. W. Moeller, President of Health Physics Society, testimony before South Carolina Legislative Investigative Committee, September 1971; "Shippingport Nuclear Power Station—Alleged Health Effects," Governor's Fact Finding Committee, State of Pennsylvania, 1974.
27. United States Code of Federal Regulations, Amendments to Title 10, pt. 50, app. I, *Federal Register* 40:19439 (May 5, 1975).
28. *Theoretical Possibilities and Consequences of Major Accidents in Large Nuclear Power Plants*, U.S. Atomic Energy Commission Report WASH-740 (Washington, D.C., 1957).
29. H. Kendall and D. Ford, "Nuclear Safety," *Environment* 14:45 (September 1972); J. Primack, "Nuclear Reactor Safety: An Introduction to the Issues," *Bull. Atomic Sci.*, September 1975, p. 15.
30. R. Nader and J. Abbots, *The Menace of Atomic Energy* (New York: Norton, 1977); E. Faltermayer, "Exorcising the Nightmare of Reactor Meltdowns," *Fortune*, March 1979, p. 82.
31. M. Maxey, "Nuclear Energy Debates: Liberation or Development," *Christian Century*, July 1976, p. 650.
32. A. M. Weinberg, "Outline for an Acceptable Nuclear Future," *Eng. & Sci.* (Cal. Inst. of Tech., Pasadena, Calif.), January 1978, p. 4.
33. Union of Concerned Scientists, *The Nuclear Fuel Cycle* (Cambridge, Mass.: M.I.T. Press, 1975); R. P. Hammond, "Nuclear Wastes and Public Acceptance," *Am. Sci.* 67:146 (1979).
34. McCracken (note 24); Ebbin and Kasper (note 24); *Energy and the Sierra Club* (San Francisco: Sierra Club, 1976).
35. King (note 23).
36. Rossin (note 14); "Atomic Industrial Forum Survey," *AIF INFO* 108:1 (July 1978).

37. See note 14.
38. See note 14.
39. King (note 23); Hosmer (note 23).
40. R. P. Feynman, "Cargo-Cult Science," *Eng. & Sci.*, September 1974, p. 15.

Chapter 3
 1. "World List of Nuclear Power Plants," *Nucl. News* February 1979, pp. 59–77; Atomic Industrial Forum, "Atomic Industrial Forum Survey of Nuclear Power Reactors, 1978," *AIF INFO* 107:1 (June 1978).
 2. National Research Council, *Report of the Uranium Resources Group Supply and Delivery Panel of the Committee on Nuclear and Alternative Energy Systems (CONAES)*, National Academy of Sciences (Washington, D.C., 1978).
 3. National Research Council (note 2); *Statistical Data of the Uranium Industry*, U.S. Department of Energy Report GJO-1100 (Washington, D.C., 1978).
 4. See note 2.
 5. *Statistical Data* (note 3).
 6. S. M. Keeny et al., *Report of the Nuclear Energy Policy Study Group: Nuclear Power Issues and Choices* (Cambridge, Mass.: Ballinger, 1977); M. A. Lieberman, "United States Uranium Reserves—An Analysis of Historical Data," *Science* 192:431 (1976).
 7. Keeny et al. (note 6).
 8. See note 2.
 9. American Physical Society, "Report to the American Physical Society by the Study Group on Light Water Reactor Safety," *Rev. Mod. Phys.* 14:546 (1975); J. Primack et al., "Nuclear Reactor Safety," *Bull. Atomic Sci.*, September 1975, pp. 15–41; R. P. Hammond, "Nuclear Power Risks," *Am. Sci.* 62:155 (1974).
10. B. L. Cohen, "Impacts of the Nuclear Energy Industry on Human Health and Safety," *Am. Sci.* 64:550 (1975); *Reactor Safety Study*, U.S. Nuclear Regulatory Commission Report WASH-1400 (Washington, D.C., 1975).
11. *Reactor Safety Study* (note 10).
12. J. D. Burtt, *LOFT Experimental Program Document*, Idaho National Engineering Laboratory Review Report, U.S. Department of Energy (Washington, D.C., December 8, 1978).
13. E. P. Alexanderson, ed., *Fermi I: New Age for Nuclear Power* (Hinsdale, Ill.: American Nuclear Society, 1979); G. Kaplan, "The Browns Ferry Incident," *IEEE Spectrum*, October 1976, p. 55.
14. Office of Nuclear Reactor Regulation, *Staff Report on the Generic Assessment of Feedwater Transients in Pressurized Water Reactors Designed by the Babcock and Wilcox Company*, U.S. Nuclear Regulatory Commission Report NUREG-0560 (Washington, D.C., May 9, 1979).

15. Dr. William Kerr, Advisory Committee on Reactor Safeguards, U.S. Nuclear Regulatory Commission, private communication, 1979.
16. D. Nelkin, "The Role of Experts in a Nuclear Siting Controversy," *Bull. Atomic Sci.* 30:29 (1974).
17. T. F. Lomenich and N. S. Stike, *Earthquakes and Nuclear Power Plant Design*, U.S. Atomic Energy Commission Report ORNL-NSIC-28 (Oak Ridge, Tenn., 1970): R. B. Matthiesen, *Earthquake Effects at Nuclear Reactor Facilities from the San Fernando Earthquake of February 9, 1971*, University of California at Los Angeles School of Engineering Report (Los Angeles, 1971).
18. J. A. Ashworth, "Off-Shore Nuclear Power Plant Siting," *Nucl. Tech.* 22:170 (1974); L. J. Carter, "Floating Nuclear Power Plants," *Science* 183:1063 (1974).
19. *Theoretical Consequences of Major Accidents in Large Nuclear Power Plants*, U.S. Atomic Energy Commission Report WASH-740 (Washington, D.C., 1957).
20. John Fuller, *We Almost Lost Detroit* (New York: Reader's Digest Publications, 1975); R. Nader and J. Abbots, *The Menace of Atomic Power* (New York: Norton, 1977); E. Faltermayer, "Exorcising the Nightmare of Reactor Meltdowns," *Fortune*, March 1979, p. 82.
21. *Reactor Safety Study* (note 10).
22. *Reactor Safety Study* (note 10).
23. H. W. Kendall and S. Moglewer, *Preliminary Review of AEC Reactor Safety Study* (San Francisco and Cambridge, Mass.: Sierra Club and Union of Concerned Scientists, 1974); N. C. Rasmussen, "The Safety Study and Its Feedback," *Bull. Atomic. Sci.*, September 1975, p. 25.
24. American Physical Society (note 9); H. W. Lewis et al., *Risk Assessment Review Group Report to the U.S.. Nuclear Regulatory Commission*, U.S. Nuclear Regulatory Commission Report NUREG/CR-0400 (Washington, D.C., 1978).
25. Extension and Phase-Out of the Price-Anderson Act of 1975, Public Law 94–197, 94th Cong., H.R. 8631, December 31, 1975.
26. "Supreme Court Upholds Price-Anderson Act," *AIF INFO* 120:1 (July 1978).
27. Ellen Thro, "Supreme Court Affirms Price-Anderson Act's Constitutionality," *Nucl. News* 21:28 (August 1978).
28. American Nuclear Society, *Nuclear Power and the Environment* (Hinsdale, Ill.: American Nuclear Society, 1976); G. G. Eichholz, *Environmental Aspects of Nuclear Power* (Ann Arbor: Ann Arbor Science, 1976), pp. 207–60.
29. F. L. Parker and P. A. Krenkl, *Physical and Engineering Aspects of Thermal Pollution* (Cleveland: CRC Press, 1970).
30. D. J. Rose, P. W. Walsh, and L. L. Leskovjan, "Nuclear Power: Compared to What?", *Am. Sci.* 64:291 (1976).
31. United States Code of Federal Regulations, Amendments to Title 10, pt. 50, app. I, *Federal Register* 40:19439 (May 5, 1975).

32. B. L. Cohen, "Impacts of the Nuclear Energy Industry on Human Health and Safety," *Am. Sci.* 64:550 (1976).

33. "Proposed Standards for Radiation Protection for Nuclear Power Operation," United States Code of Federal Regulations, Title 10, pt. 190, *Federal Register* 40:104 (May 29, 1975).

34. J. P. McBride, R. E. Moore, J. P. Witherspoon, and R. E. Blanco, "Radiological Impact of Airborne Effluents of Coal and Nuclear Plants," *Science* 202:1045 (1978); C. E. Whittle et al., *Economic and Environmental Implications of a United States Nuclear Moratorium, 1985– 2010,* Institute for Energy Analysis Report ORAU/IEA 76-4 (Oak Ridge, Tenn., 1976).

35. See note 32.

36. J. Tadmor, "Risk and Safety in the Nuclear Industry and Conventional Norms of Society," *Radioprotection* 12:275 (1977); G. H. Whipple, "Low Level Radiation: Is There a Need to Reduce the Limit?" (Paper presented at the Atomic Industrial Forum Conference on Nuclear Power: Issues and Audiences, Houston, 1978); H. Inhaber, "Risk with Energy from Conventional and Nonconventional Sources," *Science* 203:718 (1979); Aaron Wildavsky, "No Risk Is the Highest Risk of All," *Am. Sci.* 67:32 (1979).

37. D. Rossin and T. A. Rieck, "Economics of Nuclear Power," *Science* 201:582 (1978); L. F. C. Reichle, "The Economics of Nuclear Power" (Paper presented to the New York Society of Security Analysts, New York, 1975 and 1976).

38. Reichle (note 37).

39. Ebasco Corporation studies submitted to the New York State Public Service Commission, Albany, N.Y., July 1977.

40. C. L. Rudasill, "Comparing Coal and Nuclear Generating Costs," *EPRI Journal* (October 1977), p. 14.

41. *Confrontation at Seabrook* (Washington, D.C.: The Heritage Foundation, 1978).

42. Frank Graham, "The Outrageous Mr. Cherry and the Underachieving Nukes," *National Audubon,* September 1977, p. 50.

43. *National Electrical Reliability Council Annual Report,* Edison Electric Institute (New York, September 8, 1977).

44. R. H. Fischer and R. S. Palmer, "The Energy Efficiency of Electric Power Plants" (Paper presented at the Sixteenth Annual ASME Symposium on Energy Alternatives, Albuquerque, N. Mex., 1976); S. Baron, *Solar Energy: Will It Conserve Our Non-renewable Resources?* (Oradell, N.J.: Burns and Roe, Inc., 1978).

Chapter 4

1. Union of Concerned Scientists, *The Nuclear Fuel Cycle* (Cambridge, Mass.: M.I.T. Press, 1975); American Physical Society, "Report to the American Physical Society by the Study Group on Nuclear Fuel Cycles and Waste Management," *Rev. Mod. Phys.* 50:S1–S185 (January 1978); *Final Generic Environmental Statement on the Use of Recycle*

Plutonium in Mixed Oxide Fuel in Light Water Cooled Reactors (GESMO), U.S. Nuclear Regulatory Commission Report NUREG-0002 (Washington, D.C., 1976); B. L. Cohen, "Impacts of the Nuclear Energy Industry on Human Health and Safety," *Am. Sci.* 64:550 (1976).

2. E. A. Mason, "The Nuclear Fuel Cycle," in *Education and Research in the Nuclear Fuel Cycle,* ed. D. M. Elliot and L. E. Weaver (Norman, Okla.: Oklahoma University Press, 1970), pp. 15–31.

3. *Statistical Data of the Uranium Industry,* U.S. Department of Energy Report GJO-1100 (78) (Grand Junction, Colo., 1978).

4. American Physical Society (note 1).

5. W. P. Beggington, "The Reprocessing of Nuclear Fuels," *Sci. Am.* 235:30 (December 1976).

6. T. H. Pigford and K. P. Ang, "The Plutonium Fuel Cycles," *Health Physics* 29:451 (1975); T. H. Pigford and C. S. Yang, *Thorium Fuel Cycles*, U.S. Environmental Protection Agency Report EPA 68-01-1962 (Washington, D.C., 1977); B. R. Sehgal, J. A. Naser, C. Lin, and W. B. Loewenstein, "Thorium Based Fuels in Fast Breeder Reactors," *Nucl. Tech.* 35:635 (1977).

7. American Physical Society, "Report to the American Physical Society by the Study Group of Nuclear Fuel Cycles and Waste Management," *Rev. Mod. Phys.* 50 (January 1978), chap. 8.

8. Union of Concerned Scientists (note 1); Cohen (note 1).

9. D. Comey, "The Legacy of Uranium Tailings," *Bull. Atomic. Sci.*, October 1975, p. 43; B. L. Cohen, "Environmental Impacts of Nuclear Power Due to Radon Emissions," *Bull. Atomic Sci.*, February 1976, p. 61; R. O. Pohl, "Health Effects of Radon-222 from Uranium Mining," *Search* 7:350 (1976).

10. American Physical Society (note 1).

11. R. M. Fry and J. E. Cook, "Comment to Health Effects of Radon-222 from Uranium Mining," *Search* 7:350 (1976).

12. L. J. Carter, "Uranium Milling Tailings: Congress Address A Long-Neglected Problem," *Science* 202:191 (1978); R. H. Kennedy, L. J. Deal, F. F. Haywood, and W. A. Goldsmith, "Management and Control of Radioactive Wastes from Uranium Milling Operations," IAEA Conference on Nuclear Power and Its Fuel Cycle, IAEA-CN-36/479 (1977).

13. M. Benedict and T. Pigford, *Nuclear Chemical Engineering* (New York: McGraw-Hill, 1957).

14. R. H. Fischer and R. S. Palmer, "The Energy Efficiency of Electric Power Plants" (Paper presented at the Sixteenth Annual ASME Symposium on Energy Alternatives, Albuquerque, N. Mex., 1976).

15. D. Olander, "The Gas Centrifuge," *Sci. Am.* 239:37 (August 1978).

16. E. W. Becker, "Gasdynamic Nozzle Separation of Uranium Isotopes," *Prog. Nucl. Energy* 1:27 (1977).

17. M. Benedict, "Enrichment—A Critical Status Report," *Trans. Am. Nucl. Soc.* 24:8 (1976).

18. B. Snavely, *Separation of Uranium Isotopes by Laser Photochemistry*, U.S. Atomic Energy Commission Report UCRL-75725 (Livermore, Calif., 1974).

19. A. L. Hammond, "Uranium: Will There Be a Shortage or an Embarrassment of Enrichment," *Science* 192:866 (1976); A. S. Krass, "Laser Enrichment of Uranium: The Proliferation Connection," *Science* 196:721 (1977); W. D. Metz, "Laser Enrichment: Time Clarifies the Difficulty", *Science* 191:1162 (1976).

20. Hammond (note 19).

21. Cohen (note 1).

22. "IF 300 Irradiated Fuel Shipping Cask," Technical Description, General Electric Company Report (San Jose, Calif., 1975).

23. American Physical Society (note 1); *GESMO* (note 1).

24. American Physical Society (note 1).

25. Cohen (note 1); *Krypton-95 in the Atmosphere: Accumulation, Biological Significance, and Control Technology*, National Council on Radiation Protection and Measurement report no. 44 (Washington, D.C., 1975).

26. American Physical Society (note 1); S. M. Keeny et al., *Report of the Nuclear Energy Policy Study Group: Nuclear Power Issues and Choices* (Cambridge, Mass.: Ballinger, 1977).

27. *Benefit Analysis of Reprocessing and Recycling Light Water Reactor Fuel*, U.S Energy Research and Development Administration Report (Washington, D.C., December 1976).

28. B. Spinrad and E. Evans, "Using Plutonium as a Fuel," *Trans. Am. Nucl. Soc.* 24:10 (1976).

29. M. Willrich and R. K. Lester, *Radioactive Waste: Management and Regulation* (New York: The Free Press, 1977); R. P. Hammond, "Nuclear Wastes and Public Acceptance," *Am. Sci.* 67:146 (1979).

30. Ibid., and American Physical Society (note 1).

31. American Physical Society (note 1, pp. 110–11); B. L. Cohen, "The Disposal of Radioactive Wastes from Fission Reactors," *Sci. Am.* 236; 21 (January 1977); B. L. Cohen, "High Level Radioactive Waste from Light-Water Reactors," *Rev. Mod. Phys.* 49:1 (1977); C. F. Smith and W. E. Kastenberg, "On Risk Assessment of High Level Radioactive Waste Disposal," *Nucl. Eng. & Des.* 39:293 (1976).

32. Cohen, *Rev. Mod. Phys.* (note 31); American Physical Society (note 1).

33. Cohen, *Rev. Mod. Phys.* and *Sci. Am.* (note 31).

34. J. McBride, R. E. Moore, J. P. Witherspoon, and R. E. Blanco, "Radiological Impact of Airborne Effluents of Coal and Nuclear Plants," *Science* 202:1045 (1978).

35. *Generic Environmental Impact Statement on Handling and Storage of Spent Light Water Power Reactor Fuel*, U.S. Nuclear Regulatory Commission Report NUREG-0404 (Washington, D.C., 1978); *Alternatives for Managing Wastes from Reactors and Post Fission Operations in the LWR Fuel Cycle*, U.S. Energy Research and Development Administration Report ERDA-43 (Washington, D.C., 1976); *High-Level*

Radioactive Waste Management Alternatives, U.S. Energy and Research and Development Administration Report BNWL-1900 (Hanford, Wash., 1975).

36. American Physical Society (note 1).
37. American Physical Society (note 1).
38. L. J. Carter, "Nuclear Wastes: The Science of Geologic Disposal Seen As Weak," *Science* 200:1135 (1978); R. A. Kerr, "Geologic Disposal of Nuclear Wastes: Salt's Lead is Challenged," *Science* 204:603 (1979).
39. *Alternatives for Managing Wastes* (note 35); *High-Level Radioactive Waste Management Alternatives* (note 35).
40. D. Berwald and J. Duderstadt, "Preliminary Design and Neutronic Analysis of a Laser Driven Fusion Actinide Waste Burning Hybrid Reactor," *Nucl. Appl.* 42:34 (1978); H. C. Clairborne, *Neutron Induced Transmutation of High Level Radioactive Waste,* U.S. Atomic Energy Commission Report ORNL-TM-3964 (Oak Ridge, Tenn., 1972).
41. American Physical Society (note 1); *Interim Storage of Solidified High-Level Radioactive Wastes,* U.S. National Academy of Sciences, Committee on Radioactive Waste Management (Washington, D.C., 1975); *Draft Report on Nuclear Waste Management by the Interagency Review Group,* Executive Office of the President, Office of Science and Technology, October, 1978 (distributed also by the U.S. Department of Energy); *Report of the Task Force for Review of Waste Management* (draft), U.S. Department of Energy Report DOE/ER-004/D (Washington, D.C., February 1978); *Geologic Disposal of High Level Radioactive Wastes: Earth Science Perspectives,* U.S. Geological Survey, Department of the Interior (Washington, D.C., May 1978); *Management of Commercial Radioactive Nuclear Wastes: A Status Report,* Federal Energy Resources Council and the Council on Environmental Quality (Washington, D.C., May 1976).
42. D. Rossin and T. A. Rieck, "Economics of Nuclear Power," *Science* 201:582 (1978).
43. Hammond (note 29).
44. A. R. Tamplin and T. B. Cochran, "Radiation Standards for Hot Particles," *New Scientist* 66:497 (1975).
45. C. L. Comar, *Plutonium: Facts and Inferences,* Electric Power Research Institute Report EPRI EA-43-SR (Palo Alto, Calif., 1976).
46. See note 45 and J. T. Edsall, "Toxicity of Plutonium and Some Other Actinides," *Bull. Atomic. Sci.,* September 1975, p. 35; W. J. Bair, "Toxicity of Plutonium," *Adv. Radiat. Biol.* 4:41 (1978); W. S. Jee, ed., *The Health Effects of Plutonium and Radium* (Salt Lake City, Utah: J. W. Press, 1976).
47. Tamplin and Cochran (note 44).
48. W. J. Bair, "Current Status of the Hot Particle Issue," *Proceedings of the Fourth Congress of the International Radiation Protection Association* (Paris, 1977), pp. 703–10.
49. B. L. Cohen, *The Hazards in Plutonium Dispersal,* Institute of Energy Analysis Report TID-26794 (Oak Ridge, Tenn., March 1975).

50. M. Flood, "Nuclear Sabotage," *Bull. Atomic Sci.* 32:29 (October 1976); D. B. Smith and I. Waddoups, "Safeguarding Nuclear Materials and Plants," *Power Eng.* 80:36 (1976).
51. *Reactor Safety Study*, U.S. Nuclear Regulatory Commission Report WASH-1400 (Washington, D.C., 1975).
52. J. McPhee, *The Curve of Binding Energy* (New York: Ballantine, 1975); T. Taylor and M. Willrich, *Nuclear Theft: Risks and Safeguards* (Cambridge, Mass.: Ballinger, 1974).
53. T. Taylor, "Nuclear Safeguards," *Adv. Nucl. Sci. Tech.* 9:407 (1975).
54. W. Meyer, S. K. Loyalka, W. E. Nelson, and R. W. Williams, "The Homemade Nuclear Bomb Syndrome," *Nucl. Safety* 4:427 (1977).
55. *Special Safeguards Study*, U.S. Nuclear Regulatory Commission Report NUREG-75-060 (Washington, D.C., 1975).

Chapter 5

1. Atomic Industrial Forum, "Atomic Industrial Forum Survey of Nuclear Power Reactors," *AIF INFO* 107:1 (June 1978); "World List of Nuclear Power Plants," *Nucl. News* 22:59–77 (February 1979).
2. *Energy: Global Prospects 1985–2000*, Workshop on Alternative Energy Strategies (New York: McGraw-Hill, 1977).
3. B. A. Hutchins, "Nuclear Programs in Other Nations" (Paper presented at the AUA-ANL Conference on International Aspects of Nuclear Power, Argonne National Laboratory, Argonne, Ill., May 16, 1978).
4. J. Walsh, "Nuclear Exports and Proliferation: The French Think They Have a Case," *Science* 193:387 (1977); F. Lewis, "A Case Study of One Nuclear Deal: France and Pakistan," *New York Times,* November 14, 1976, p. E22.
5. N. Gall, "Atoms for Brazil: Dangers for All," *Bull. Atomic Sci.,* June 1975, p. 5; A. L. Hammond, "Brazil's Nuclear Program: Carter's Nonproliferation Policy Backfires," *Science* 193:657 (1977).
6. C. Hinton, "Atomic Power in Britain," *Sci. Am.* 198:29 (March 1958); K. P. Gibbs and D. R. Fair, "The Magnox Stations: A Success Story," *Nucleonics* 24:43 (September 1966).
7. S. Rippon, "Capenhurst Centrifuge Plant Inaugurated," *Nucl. News* 20:54 (November 1977).
8. G. A. Vendreyes, "Superphenix: A Full-Scale Breeder Reactor," *Sci. Am.* 236:26 (1976); S. Rippon, "Super Progress on Superphenix," *Nucl. News* 22:63 (March 1979).
9. H. C. McIntyre, "Natural-Uranium Heavy-Water Reactors," *Sci. Am.* 233:17 (1975); J. A. L. Robinson, "The CANDU Reactor," *Science* 199:657 (1978).
10. S. M. Keeny et al., *Report of the Nuclear Energy Policy Study Group: Nuclear Power Issues and Choices* (Cambridge, Mass.: Ballinger, 1977); letter from S. E. Eizenstat to Dr. J. M. Hendrie, Chairman, U.S. Nuclear Regulatory Commission, October 4, 1977; W. D. Metz, "Car-

ter's New Plutonium Policy: Maybe Less Than Meets the Eye," *Science* 196:405 (1977); Jimmy Carter, "Three Steps Toward Nuclear Responsibility," *Bull. Atomic Sci.,* October 1976, p. 8.

11. S. Rippon, "Comment from Europe: Politeness Giving Way to Resentment," *Nucl. News* 21:68 (October 1978); C. Starr, "Nuclear Power and Weapons Proliferation—the Thin Link," *Nucl. News* 20:54 (June 1977).
12. N. Hawkes, "Science in Europe: The Antinuclear Movement Takes Hold," *Science* 197:1167 (1977).
13. M. Rosen, "The Critical Issue of Nuclear Power Plant Safety in Developing Countries," *IAEA Bull.* 19:12 (1977).
14. "Technology Transfer," *General Electric Nuclear Power Newsletter*, Nuclear Energy Division, General Electric Company (San Jose, Calif., Fall 1977).
15. See note 13.
16. Ibid.
17. F. Ilke, "Illusions and Realities about Nuclear Energy," *Bull. Atomic Sci.*, October 1976, p. 15; B. Feld, "Nuclear Proliferation—Thirty Years After Hiroshima," *Phys. Today*, July 1975, p. 23; W. Epstein, "The Proliferation of Nuclear Weapons," *Sci. Am.* 233:18 (April 1975).
18. H. A. Feiverson and T. B. Taylor, *Alternative Strategies for International Control of Nuclear Power* (Cambridge, Mass.: Ballinger, 1977).
19. B. Goldschmidt, "A Historical Survey of Nonproliferation Policies," *Int. Security* 2:69 (1977).
20. Metz (note 10); Eizenstat letter (note 10).
21. Starr (note 11); Goldschmidt (note 19); E. L. Zebroski, "International Thermal Reactor Development," *Proceedings of the Sixteenth Nuclear Engineering Education Conference on International Nuclear Engineering*, Argonne National Laboratory (Argonne, Ill., March 1977); C. Walske, "Civilian Nuclear Power Without Weapons Proliferation" (Paper presented at Fuel Cycle Conference, New York, March 1978); D. J. Rose and R. K. Lester, "Nuclear Power, Nuclear Weapons, and International Stability," *Sci. Am.* 238:45 (April 1978).
22. See note 5.
23. Ibid.
24. Rose and Lester (note 21); *Nuclear Proliferation and Safeguards*, Office of Technology Assessment, Congress of the United States (New York: Praeger, 1977); *Nuclear Power and Nuclear Weapons Proliferation*, Report of the Atlantic Council's Nuclear Fuels Policy Working Group, vol. 1 (Washington, D.C.: The Atlantic Council, 1978); M. B. Kratzer and B. M. Jones, "Nuclear Power and Weapons Proliferation—an Optimistic View," *Nucl. News* 21:67 (October 1978).
25. T. Taylor, "Nuclear Safeguards," *Adv. Nucl. Sci. Tech.* 9:407 (1975); E. J. Moniz and T. L. Neff, "Nuclear Power and Nuclear Weapons Proliferation," *Phys. Today*, April 1978, p. 42.
26. Epstein (note 17).

27. American Physical Society, "Report to the American Physical Society by the Study Group on Nuclear Fuel Cycles and Waste Management," *Rev. Mod. Phys.* 50:S1–S185 (1978).

28. W. D. Metz, "Laser Enrichment: Time Clarifies the Difficulty," *Science* 191:1162 (1976); A. S. Krass, "Laser Enrichment of Uranium: The Proliferation Connection," *Science* 196:721 (1977).

29. J. J. Glackin, "Nuclear Proliferation," letter to the editor, *Science* 189:944 (1977).

30. J. R. Lamarsh, "On the Construction of Pu-Producing Reactors by Small and/or Developing Nations" (Paper prepared for the Congressional Research Service of the Library of Congress, April 30, 1976).

31. W. Meyer, S. K. Loyalka, W. E. Nelsen, and R. W. Williams, "The Homemade Nuclear Bomb Syndrome," *Nucl. Safety* 4:427 (1977).

32. See note 30.

33. See note 4.

34. Metz (note 10); Eizenstat letter (note 10); Carter (note 10).

35. Rippon (note 11).

36. See note 18; W. D. Metz, "Reprocessing Alternatives: The Options Multiply," *Science* 196:284 (1977); K. Cohen, "The Science and Science Fiction of Reprocessing and Proliferation" (Paper presented at the Nuclear Fuel Cycle Conference, Kansas City, Mo., 1977).

37. See note 34.

38. Metz (note 36); Cohen (note 36).

39. Cohen (note 36); American Physical Society (note 27).

40. Cohen (note 36); E. E. Till, E. M. Bohn, Y. I. Chang, and J. B. van Erp, *A Survey of Considerations Involved in Introducing CANDU Reactors into the United States*, U.S. Energy Research and Development Administration Report ANL-76-132 (Argonne, Ill., 1977).

41. H. A. Feiverson, F. von Hippel, and R. H. Williams, "Fission Power: An Evolutionary Strategy," *Science* 203:330 (1979).

42. See note 27 and Cohen (note 36).

43. C. Starr, "The Separation of Nuclear Power from Nuclear Proliferation" (Paper presented at the Fifth Energy Technology Conference, Washington, D.C., February 27, 1978); "CIVEX: Solution to Breeder-Diversion Dilemma?", *Nucl. News* 21:32 (April 1978).

44. Cohen (note 36).

45. Zebroski (note 21).

46. B. Goldschmidt (note 19).

47. W. Epstein (note 17).

48. D. B. Smith and I. Waddoups, "Safeguarding Nuclear Materials and Plants," *Power Eng.*, November 1976, p. 36.

49. Goldschmidt (note 19); Epstein (note 17).

50. See note 10.

Chapter 6

1. G. A. Vendreyes, "Superphenix: A Full-Scale Breeder Reactor," *Sci. Am.* 236:26 (March 1977); W. Hafele and C. Starr, "The Liquid Metal

Fast Breeder Reactor," *J. Brit. Nucl. Energy Soc.* 13:131 (1974); U.S., Congress, Joint Committee on Atomic Energy, *Issues for Consideration—Review of the National Breeder Reactor Program,* 94th Cong., 1st sess. August 1975.

2. W. H. Hannum and J. D. Griffith, "Reactors—Safe at Any Speed," *Trans. Am. Nucl. Soc.* 22:283 (1975).

3. S. M. Keeny et al., *Report of the Nuclear Energy Policy Study Group: Nuclear Power Issues and Choices* (Cambridge, Mass.: Ballinger, 1977); H. Bethe, "The Necessity of Fission Power," *Sci. Am.* 234:1 (January 1975).

4. "LWBR Goes Critical," *Nucl. News* 20:35 (October 1977).

5. See note 1 and M. Banal et al., "Creys-Malville Nuclear Power Station," *Nucl. Eng. Int.*, June 1978, pp. 43–60; S. Rippon, "Super Progress on Superphenix," *Nucl. News* 22:63 (March 1979).

6. T. Alexander, "Why the Breeder Reactor Is Inevitable," *Fortune*, September 1977, p. 123.

7. *Clinch River Breeder Reactor Plant Project: Design Description*, (Oak Ridge, Tenn.: Breeder Reactor Corporation, January 1978).

8. Keeny et al. (note 3); T. R. Stauffer, H. L. Wyckoff, and R. S. Palmer, "An Assessment of Economic Incentives for the LMFBR" (Paper presented to the Breeder Reactor Corporation, Chicago, Ill., March 1975); M. Levenson, P. M. Murphy, and C. P. Zaleski, "Economic Perspective of the LMFBR," *Nucl. News* 19:54 (April 1976).

9. Stauffer, Wyckoff, and Palmer (note 8).

10. R. F. Post and F. L. Ribe, "Fusion Power," *Science* 186:397 (1974); R. F. Post, "Nuclear Fusion," *Ann. Rev. Energy* 1:213 (1976); D. Rose and M. Feiertag, "Fusion Power," *Tech. Rev.* 79:20 (1976).

11. Post (note 10).

12. T. B. Cochran, *The Liquid Metal Fast Breeder Reactor: An Environmental and Economic Critique* (Baltimore: John Hopkins University Press, 1974); T. B. Cochran, J. G. Speth, and A. R. Tamplin, "Bypassing the Breeder," National Resources Defense Council Report (Washington, D.C., March 1975).

13. Amasa S. Bishop, *Project Sherwood: The U.S. Program in Controlled Fusion* (New York: Doubleday, Anchor Books, 1960).

14. Post (note 10).

15. G. L. Kulcinski, *Critical Issues Facing the Long-Term Deployment of Fission and Fusion Breeder Reactors*, University of Wisconsin Report UWFDM-234 (Madison, Wis., March 1978); G. L. Kulcinski, G. Kessler, J. P. Holdren, and W. Hafele, "Energy for the Long Run: Fission or Fusion?", *Am. Sci.* 67:78 (1979); John P. Holdren, "Fusion Energy in Context: Its Fitness for the Long Term," *Science* 200:168 (1978).

16. Post (note 10); D. Rose, "Controlled Nuclear Fusion: Status and Outlook," *Science* 172:797 (1971); S. Glasstone and R. H. Lovberg, *Controlled Thermonuclear Reactions* (New York: Van Nostrand, 1960).

17. Post (note 10); Glasstone and Lovberg (note 16).

18. Post (note 10).

19. Post (note 10).
20. F. L. Ribe, "Fusion Power Systems," *Trans. Am. Nucl. Soc.* 24:32 (1976); F. L. Ribe, *Recent Developments in the Design of Conceptual Fusion Reactors*, Los Alamos Scientific Laboratory Report LA UR-76-2165 (1976).
21. D. Steiner, "The Technological Requirements for Power by Fusion," *Nucl. Sci. Eng.* 58:107 (1975).
22. E. E. Kintner, "Status of the Magnetic Fusion Program," U.S. Department of Energy Program Review, June 14, 1978.
23. B. Badger et al., *UWMAK-I, the Wisconsin Tokamak Reactor Design*, U.S. Atomic Energy Commission Report CONF-470402 (Washington, D.C., 1974), p. 38; R. G. Mills, *A Tokamak Fusion Power Reactor*, U.S. Atomic Energy Commission Report MATT-1050 (Princeton, N.J., 1974); W. D. Metz, "Fusion Research I: What Is the Program Buying the Country," *Science* 192:1320 (1976); W. D. Metz, "Fusion Research II: Detailed Reactor Studies Identify More Problems," *Science* 193:38 (1976); P. H. Abelson, "Glamorous Nuclear Fusion," *Science* 193:5 (1976).
24. Holdren (note 15); J. P. Holdren, *Safety and Environmental Aspects of Fusion Power Plants*, U.S. Energy Research and Development Administration Report UCRL-78759 (Livermore, Calif., 1976); D. Okrent et al., "On the Safety of Tokamak Type Central Station Fusion Power Reactors," *Nucl. Eng. Des.* 39:215 (1976).
25. Holdren, U.S.E.R.D.A. Report (note 24).
26. Steiner (note 21).
27. Holdren (note 15); D. Steiner and J. F. Clarke, "The Tokamak: Model-T Fusion Reactor," *Science* 199:1395 (1978).
28. W. E. Parkins, "Engineering Limitations of Fusion Power Plants," *Science* 199:1403 (1978).
29. Steiner and Clarke (note 27); G. L. Kulcinski and C. W. Maynard, "NUWMAK: An Attractive Medium Field, Medium Size, Conceptual Tokamak Reactor," *Proceedings of Third Topical Meeting on Technology of Controlled Fusion*, American Nuclear Society (Santa Fe, N. Mex., May 1978); G. H. Miley and J. G. Gilligan, "A Possible Route to Small, Flexible Fusion Units" (Paper presented at Midwest Energy Conference, Chicago, Ill., November 1978).
30. Kulcinski (note 15).
31. Kulcinski et al. (note 15); also see note 28.
32. Kulcinski et al. (note 15).
33. Metz (note 23); Abelson (note 23).
34. B. R. Leonard, "A Review of Fission-Fusion Hybrid Concepts," *Nucl. Tech.* 20:161 (1973); L. Lidsky, "Fission-Fusion Systems: Hybrid, Symbiotic, and Augean," *Nucl. Fusion* 15:151 (1975); W. D. Metz, "Fusion Research III: New Interest in Fusion Assisted Breeders," *Science* 193:307 (1976).
35. D. R. Berwald and J. J. Duderstadt, "Preliminary Design and Neutronic

Analysis of a Laser Fusion Driven Actinide Waste Burning Hybrid Reactor," *Nucl. Tech.* 42:34 (1978).

36. E. Teller, "A Future ICE (Thermonuclear, That Is!)," *IEEE Spectrum*, January 1973, p. 60.

37. J. Nuckolls, L. Wood, A. Thiessen, and G. Zimmerman, "Laser Compression of Matter to Super-High Densities: Thermonuclear (CTR) Applications," *Nature* 239:139 (1972); J. L. Emmett, J. Nuckolls, and L. Wood, "Fusion Power by Laser Implosion," *Sci. Am.* 231:24 (June 1974); K. Brueckner and S. Jorna, "Laser Driven Fusion," *Rev. Mod. Phys.* 46:325 (1974); J. S. Clarke, H. N. Fisher, and R. J. Mason, "Laser Driven Implosion of Spherical DT Targets to Thermonuclear Burn Conditions," *Phys. Rev. Lett.* 30:89 (1974).

38. A. Fraas and M. Lubin, "Fusion by Laser," *Sci. Am.* 225:21 (June 1971).

39. K. Boyer, "Laser Driven Fusion," *Aero and Astro*, July 1973, p. 28.

40. P. M. Campbell, G. Charatis, and G. R. Montry, "Laser Driven Compressions of Glass Microspheres," *Phys. Rev. Lett.* 34:74 (1975).

41. T. J. Burgess, "Lasers for Fusion Systems," *IEEE Trans. Plas. Sci.* 1:26 (1973); J. Wilson and D. O. Ham, "Brand-X Lasers for Laser Fusion," *Laser Focus* 12:38 (1976).

42. G. Yonas, "Fusion Power with Particle Beams," *Sci. Am.* 239:48 (November 1978); W. D. Metz, "Energy Research: Accelerator Builders Eager to Aid Fusion Work," *Science* 194:307 (1976).

43. L. A. Booth, D. A. Freiwald, T. G. Frank, and F. T. Finch, "A Laser Fusion Reactor Design," *Proc. IEEE* 64:1460 (1976).

44. Kulcinski et al. (note 15); Holdren (note 15); Metz (note 23); Abelson (note 23).

45. *Final Report of the Ad Hoc Experts Group on Fusion* (The Foster Committee), U.S. Department of Energy Report DOE/ER-0008 (Washington, D.C., June 1978).

46. Holdren (note 15).

47. Abelson (note 23).

Selected Bibliography

General Review

Bethe, H. "The Necessity of Fission Power." *Sci. Am.* 234:5 (January 1975).

Campana, R. J., and Langer, S. *Q & A: Nuclear Power and the Environment.* Hinsdale, Ill.: American Nuclear Society, 1976.

Cohen, B. L. *Nuclear Science and Society.* Garden City, N.J.: Doubleday, Anchor Books, 1974.

Keeny, S. M., et al. *Report of the Nuclear Energy Policy Study Group: Nuclear Power Issues and Choices.* Cambridge, Mass.: Ballinger, 1977.

National Research Council. *Report of the Committee on Nuclear and Alternative Energy Systems* CONAES). National Academy of Sciences, Washington, D.C., 1979.

Schmidt, F. H., and Bodansky, D. *The Fight over Nuclear Power.* San Francisco: Albion, 1976.

Weaver, K. F. "The Promise and Peril of Nuclear Energy." *Nat. Geo.* 155:459 (April 1979).

Whittle, C. E., et al. *Economic and Environmental Implications of a United States Nuclear Moratorium, 1985–2010.* Institute for Energy Analysis Report ORAU/IEA 76–4. Oak Ridge, Tenn., 1976.

Technical Aspects

Connolly, T. M. *Foundations of Nuclear Engineering.* New York: Wiley, 1978.

Duderstadt, J. J. *Nuclear Power.* New York: Marcel Dekker, 1979.

Glasstone, S., and Sesonske, A. *Nuclear Reactor Engineering.* 2d ed. Princeton, N.J.: Van Nostrand, 1977.

Lamarsh, J. R. *Introduction to Nuclear Engineering.* Reading, Mass.: Addison-Wesley, 1975.

Murray, R. L. *Nuclear Energy.* New York: Pergamon, 1975.

History of Nuclear Power

Bishop, A. S. *Project Sherwood: The U.S. Program in Controlled Fusion.* New York: Doubleday, Anchor Books, 1960.

Bupp, Irwin C. *Light Water: How the Nuclear Dream Dissolved.* New York: Derian, 1978.

Goldschmidt, B. "A Historical Survey of Nonproliferation Policies." *Int. Security* 2:69 (1977).

Groueff, Stephanie. *Manhattan Project.* Boston: Little & Brown, 1967.

Hewlett, R. G. *The Atomic Shield.* University Park, Pa.: Pennsylvania State University Press, 1972.

Hewlett, R. G., and Anderson, O. E., Jr. *The New World, 1939/1946.* University Park, Pa.: Pennsylvania State University, 1962.

Smyth, Henry D. *Atomic Energy for Military Purposes.* Princeton, N.J.: Princeton University Press, 1945.

Status, Economics, and Reliability of Nuclear Power

American Nuclear Society. "World List of Nuclear Power Plants." *Nucl. News* 22:59–77 (February 1979).

Atomic Industrial Forum. "Atomic Industrial Forum Annual Survey on World Nuclear Power Capacity." *AIF INFO* 107:1 (June 1978).

Rombaugh, C. T., and Koen, B. V. "Total Energy Investment in Nuclear Power Plants." *Nucl. Tech.* 26:5 (1975).

Rossin, D., and Rieck, T. A. "Nuclear Power Economics." *Science* 201:582 (1978).

Update: Nuclear Power Program Information and Data, June 1978. Division of Nuclear Power Development, U.S. Department of Energy. Washington, D.C., June 1978.

Nuclear Power, Radiation, and the Environment

Cohen, B. L. "Impacts of the Nuclear Energy Industry on Human Health and Safety." *Am. Sci.* 64:550 (1976).

Eichholz, G. G. *Environmental Aspects of Nuclear Power.* Ann Arbor: Ann Arbor Science, 1976.

McBride, J. P.; Moore, R. E.; Witherspoon, J. P.; and Blanco, R. E. "Radiological Impact of Airborne Effluents of Coal and Nuclear Plants." *Science* 202:1045 (1978).

National Research Council, *Report of the Committee on Biological Effects of Ionizing Radiation (BEIR),* National Academy of Sciences, Washington, D.C., 1979.

Rose, D. J.; Walsh, P. W.; and Leskovjan, L. L. "Nuclear Power: Compared to What?" *Am. Sci.* 64:291 (1976).

Tadmor, J. "Risk and Safety in the Nuclear Industry and Conventional Norms of Society." *Radioprotection* 12:275 (1977).

Risk and Nuclear Reactor Safety

American Physical Society. "Report to the American Physical Society by the Study Group on Light Water Reactor Safety." *Rev. Mod. Phys.* 14:546 (1976).

Cohen, B. L. "Impacts of the Nuclear Energy Industry on Human Health and Safety." *Am. Sci.* 64:550 (1976).

Hammond, R. P. "Nuclear Power Risks." *Am. Sci.* 62:155 (1974).

Inhaber, H. "Risk with Energy from Conventional and Nonconventional Sources." *Science* 203:718 (1979).

Reactor Safety Study. U.S. Nuclear Regulatory Commission Report WASH-1400. Washington, D.C., 1975.

Wildavsky, Aaron. "No Risk Is the Highest Risk of All." *Am. Sci.* 67:32 (1979).

The Nuclear Fuel Cycle
American Physical Society. "Report to the American Physical Society by the Study Group on Nuclear Fuel Cycles and Waste Management." *Rev. Mod. Phys.* 50:S1–S185 (1978).
Beggington, W. P. "The Reprocessing of Nuclear Fuels." *Sci. Am.* 235:30 (December 1976).
Final Generic Environmental Statement on the Use of Recycle Plutonium in Mixed Oxide Fuel in Light Water Cooled Reactors (GESMO). U.S. Nuclear Regulatory Commission Report NUREG-0002. Washington, D.C., August 1976.
Graves, H., Jr. *Nuclear Fuel Management.* New York: Wiley, 1978.
Holdren, J. P. "Hazards of the Nuclear Fuel Cycle." *Bull. Atomic Sci.* 30:14 (October 1974).
Pigford, T. H. "The Nuclear Fuel Cycle." *Ann. Rev. Nucl. Sci.* 8:515 (1974).

Radioactive Waste
American Physical Society. "Report to the American Physical Society by the Study Group on Nuclear Fuel Cycles and Waste Management." *Rev. Mod. Phys.* 50:S1–S185 (1978).
Cohen, B. L. "The Disposal of Radioactive Wastes from Fission Reactors." *Sci. Am.* 236:21 (January 1977).
———. "High Level Radioactive Waste from Light-Water Reactors." *Rev. Mod. Phys.* 49:1 (1977).
Generic Environmental Impact Statement on Handling and Storage of Spent Light Water Power Reactor Fuel. U.S. Nuclear Regulatory Commission Report NUREG-0404. Washington, D.C., 1978.
Hammond, R. P. "Nuclear Wastes and Public Acceptance." *Am. Sci.* 67:146 (1979).

Plutonium
Cohen, B. L. "Impacts of the Nuclear Energy Industry on Human Health and Safety," *Am. Sci.* 64:550 (1976).
Comar, C. L. *Plutonium: Facts and Inferences.* Electric Power Research Institute Report EPRI EA-43-SR. Palo Alto, Calif., 1976.
Meyer, W.; Loyalka, S. K.; Nelson, W. E.; and Williams, R. W. "The Homemade Nuclear Bomb Syndrome." *Nucl. Safety* 4:427 (1977).
Taylor, T. "Nuclear Safeguards." *Adv. Nucl. Sci. Tech.* 9:407 (1975).
Taylor, T., and Willrich, M. *Nuclear Theft: Risks and Safeguards.* Cambridge, Mass.: Ballinger, 1974.

International Aspects of Nuclear Power
American Nuclear Society. "World List of Nuclear Power Plants." *Nucl. News* 22:59–77 (February 1979).

Atomic Industrial Forum. "Atomic Industrial Forum Survey of World Nuclear Power Reactors, 1978." *AIF INFO* 107:1 (June 1978).
Epstein, W. "The Proliferation of Nuclear Weapons." *Sci. Am.* 233:18 (April 1975).
Nuclear Power and Nuclear Weapons Proliferation. Report of the Atlantic Council's Nuclear Fuels Policy Working Group, vol. 1. Washington, D.C.: The Atlantic Council, 1978.
Rose, D. J., and Lester, R. K. "Nuclear Power, Nuclear Weapons, and International Stability." *Sci. Am.* 238:45 (April 1978).
Starr, C. "Nuclear Power and Weapons Proliferation—the Thin Link." *Nucl. News* 20:54 (June 1977).

Breeder Reactors
Alexander, T. "Why the Breeder Reactor Is Inevitable." *Fortune*, September 1977, p. 123.
Banal, M., et al. "Creys-Malville Nuclear Power Station." *Nucl. Eng. Int.*, June 1978, pp. 43–60.
Vendryes, Georges A. "Superphenix: A Full-Scale Breeder Reactor." *Sci. Am.* 236:26 (March 1977).

Nuclear Fusion
Emmett, J. L.; Nuckolls, J.; and Wood, L. "Fusion Power by Laser Implosion." *Sci. Am.* 231:24 (June 1974).
Holdren, John P. "Fusion Energy in Context: Its Fitness for the Long Term." *Science* 200:168 (1978).
Kulcinski, G. L.; Kessler, G.; Holdren, J.; and Hafele, W. "Energy for the Long Run: Fission or Fusion." *Am. Sci.* 67:78 (1979).
Post, R. F. "Nuclear Fusion." *Ann. Rev. Energy* 1:213 (1976).
Rose, D., and Feiertag, M. "Fusion Power." *Tech. Rev.* 79:20 (1976).
Steiner, D., and Clarke, J. F. "The Tokamak: Model-T Fusion Reactor." *Science* 199:1395 (1978).
Yonas, G. "Fusion Power with Particle Beams." *Sci. Am.* 239:48 (November 1978).

Alternative Energy Sources
Kruger, P. "Geothermal Energy." *Ann. Rev. Energy* 1:159 (1976).
Metz, W. D., and Hammond, A. L. *Solar Energy in America*. Washington, D.C.: American Association for the Advancement of Science, 1978.
Pollard, W. G. "The Long Range Prospects for Solar Derived Fuels." *Am. Sci.* 64:509 (1976).
———. "Solar Power." *Am. Sci.* 64:424 (1976).
Putnam, P. C. *Energy in the Future*. New York: Van Nostrand, 1953.

The Energy Crisis
Brown, Harrison. "Energy in Our Future." *Ann. Rev. Energy* 1:1 (1976).
Flower, A. R. "World Oil Production." *Sci. Am.* 238:42 (March 1978).
Fowler, John M. *Energy and the Environment*. New York: McGraw-Hill, 1975.

Griffith, E. D., and Clarke, A. W. "World Coal Production." *Sci. Am.* 240: 38 (January 1979).

Hayes, Earl T. "Energy Resources Available to the United States, 1985–2000." *Science* 203:233 (1979).

Hubbert, M. King. "The Energy Resources of the Earth." *Sci. Am.* 224:60 (September 1971).

Lapp, R. *The Logarithmic Century.* Englewood Cliffs, N.J.: Prentice-Hall, 1973.

Putnam, P. C. *Energy in the Future.* New York: Van Nostrand, 1953.

Ruedisili, L. C., and Firebaugh, M. W., eds. *Perspectives on Energy.* 2d ed. New York: Oxford University Press, 1977.

Weinberg, A. M. "Reflections on the Energy Wars." *Am. Sci.* 66:153 (1978).

Wilson, Richard, and Jones, W. J. *Energy, Ecology, and the Environment.* New York: Academic Press, 1974.

The Opposition to Nuclear Power

Berger, J. J. *Nuclear Power: The Unviable Option.* Palo Alto, Calif.: Ramparts, 1976.

Ebbin, S., and Kasper, R. *Citizen Groups and the Nuclear Power Controversy.* Cambridge, Mass.: M.I.T. Press, 1974.

Faltermayer, E. "It is Time to End the Holy War over Nuclear Power." *Fortune*, March 1979, p. 81.

Fuller, John. *We Almost Lost Detroit.* New York: Readers Digest Publications, 1975.

Gofman, J., and Tamplin, A. R. *Poisoned Power.* Emmaus, Pa.: Rodale Press, 1971.

Inglis, D. R. *Nuclear Energy: Its Physics and Its Social Challenge.* Reading, Mass.: Addison-Wesley, 1973.

Lewis, R. S. *The Nuclear Power Rebellion: Citizens v. the Atomic Industrial Establishment.* New York: Viking, 1972.

Lovins, Amory B. *Soft Energy Paths: Toward a Durable Peace.* Cambridge, Mass.: Ballinger, 1977.

McCracken, S. "The War Against the Atom." *Commentary*, August 1977, p. 33.

Nader, R., and Abbotts, J. *The Menace of Atomic Energy.* New York: Norton, 1977.

Novick, Sheldon. *The Careless Atom.* Boston: Houghton Mifflin, 1969.

———. *The Electric War.* San Francisco: Sierra Club, 1976.

Union of Concerned Scientists. *The Nuclear Fuel Cycle.* Cambridge, Mass.: M.I.T. Press, 1975.

Index